D1582644

50

IDEAS YOU REALLY NEED TO KNOW

SCIENCE

PAUL PARSONS & GAIL DIXON

Quercus

Contents

Introduction 3

Introduction

Albert Einstein once remarked: 'The greatest scientists are artists as well.' It was a bold thing to say, because science can appear a very uncreative process. Constrained by data, facts and evidence, it might seem as if there's little room for the power of ideas, and the power of creativity. But in fact, nothing could be further from the truth. Einstein's point was that the truly pioneering minds in science are also some of the most creative. And it's their ideas rather than their technical ability that have changed the world.

It's true that the vast majority of scientific research builds upon the work of others – science generally is a gradual evolution in our understanding. But every so often, a scientist of prodigious creative ability comes to the fore, and their insights bring not evolution but revolution – often turning their field on its head, and taking our understanding to a whole new level. Take, for example, Einstein and his elegant theory of relativity, Darwin and his radical ideas on evolution by natural selection, or Richard Feynman's bold re-imagining of the subatomic particle world.

Not that these individuals didn't possess great levels of technical skill as well – of course, they did. But without that spark of creative genius, no amount of mathematical aptitude or knowledge of the natural world could have brought about such wholesale transformations in science. It's a useful reminder for educators that science in school shouldn't all be about rote learning and passing tests.

Laid out in the chapters ahead are 50 of the greatest ideas that scientists have come up with over the ages. The choice of topics is quite personal – doubtless, if someone else had written this book, the contents page would look very different. But we've attempted to keep a good balance, and we hope you find them as interesting as we do. Where space permits, we've supplemented the science with brief biographies of the scientists themselves, detailing their backgrounds and personal lives. It's a fascinating journey into the accomplishments of some of science's most creative minds. We can only imagine what exciting new ideas are fomenting in the minds and on the drawing boards of scientists today – and, indeed, over the years to come.

Paul Parsons & Gail Dixon

01 Fermat's principle

In the late 17th century, French mathematician Pierre de Fermat summed up the behaviour of light rays in a single, beautiful law: light travelling between two points takes the quickest possible path. His idea would pave the way for an even more powerful principle that sits at the heart of modern theoretical physics.

By 1662, physicists had long known about the phenomenon of 'refraction', the way a beam of light bends sharply when crossing the interface between one substance and another. A good example is air and water: dip a pencil into a glass of water and, when viewed from the side, the pencil appears bent at an impossible angle. Refraction occurs when the two substances forming the interface have different 'optical densities', making the light travel at different speeds within them: the angle by which the beam is bent can be calculated by plugging the ratio of these speeds into a mathematical formula called Snell's law (see box, opposite). What wasn't so clear was why this should be so.

FERMAT'S PREVIOUS THEOREM
Fermat's insight was to propose what he called the 'principle of least time'. In essence, light would always take the quickest route between two points. This fundamental assumption, plus a little maths, led directly to Snell's law.

A good analogy is a lifeguard on a beach trying to reach a swimmer in distress. The lifeguard is some way along the beach from the swimmer, meaning that a combination of running along the beach and swimming

TIMELINE

984 CE	1622	1744
Arabic mathematician Ibn Sahl publishes the first known instance of Snell's law	Pierre de Fermat suggests that light rays follow the principle of least time	French mathematician Pierre-Louis Maupertuis proposes the principle of least action

through the water is required. The question is: how much of each?

The shortest distance between the lifeguard and the swimmer is just a straight line connecting the two, so you might expect that also to be the quickest route. But it's not, because the lifeguard can run much faster than she can swim. Taking the straight-line path would mean spending far too much time moving slowly through the water. Running along the beach to get as close as possible to the swimmer before diving in also isn't optimal, because the total distance travelled is just too far. The quickest route is a balance between the two: running diagonally down the beach towards a carefully calculated point on the shoreline, then turning sharply and heading directly towards the swimmer through the water – just like a ray of light being refracted.

The physical justification for Fermat's principle comes from the wave theory of light, in particular the phenomenon of interference – the way two waves combine into one. If the peaks of one wave coincide with the troughs of the other they will cancel out – which is called destructive interference. On the other hand, if peaks and troughs line up the result is one very big wave – constructive interference. For almost every possible path of the light ray, there will be another that interferes destructively to cancel it out. The exception is the

Snell's law

Although named after the Dutch mathematician Willebrord Snellius, the first account of what's today known as Snell's law dates back more than 600 years earlier, to the work of the Muslim mathematician Ibn Sahl.

Given an interface between two media in which the speed of light is v_1 and v_2, then the respective angles, θ_1 and θ_2, between the light beams and a line perpendicular to the interface is given by the formula

$$\sin \theta_1 / \sin \theta_2 = v_1 / v_2$$

(where sin is the standard trigonometric function).

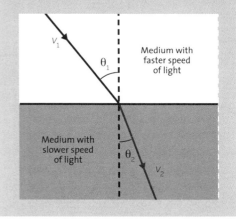

1788
Joseph-Louis Lagrange uses Maupertuis's principle to develop Lagrangian dynamics

1915
German mathematician David Hilbert derives Einstein's general relativity from an action

1948
American physicist Richard Feynman formulates the path integral approach to quantum theory

Lagrangian dynamics

Analysing the behaviour of complex dynamical systems was nightmarishly complicated until Italian mathematician Joseph-Louis Lagrange used the principle of least action to come up with what's now known as Lagrangian dynamics.

It offered a systematic way of solving problems involving the motion of many objects under the action of different forces – a good example being the Sun and the many bodies orbiting in the Solar System, all moving in response to the gravitational interactions between each other.

Lagrange's approach involved setting up coordinates for the position and speed of every object. This enabled him to write down a general form for the Lagrangian – the total kinetic energy minus potential energy of the system. Summing this over all possible paths through space and time yielded a formula for the 'action' (see page 7), which Lagrange then minimized to give a generalized set of equations describing the motion of each object.

In the case of the Solar System, the relative speeds of the Sun and the planets implied the total kinetic energy, while the position of every object relative to every other gave the total potential energy, which in this case is supplied by gravity. The resulting equations of motion describe the orbits of the planets around the Sun.

one path that minimizes the travel time, which is why that's the path that we see the light ray take. Fermat's principle also explained the laws governing how light is reflected at an interface between two media, including the phenomenon of 'total internal reflection', in which a light beam in a dense medium is unable to escape when it strikes the surface at a very shallow 'grazing' angle. This idea is key to how fibre optic cables work.

But there were much bigger developments to come. In 1744, another French mathematician, Pierre-Louis Maupertuis, wondered whether Fermat's principle could be extended to explain not just the behaviour of light rays but also the dynamics of moving objects. He replaced time in Fermat's reasoning with the kinetic energy of an object added up along the particular path that it travels. Then he postulated that the path the object actually takes is the one for which this quantity is minimized.

ACTIONS SPEAK LOUDEST

Later that century, the idea was refined further by Italian mathematician Joseph-Louis Lagrange and Irish physicist William Rowan Hamilton. They tweaked Maupertuis's theorem to sum not just the kinetic energy of a moving object, but the kinetic energy minus its stored or 'potential' energy. (For example, a stone fired from a catapult will start out with zero kinetic energy, but a large potential energy stored in the stretched rubber.) Lagrange and Hamilton argued that it was this new summed quantity, known as the 'action', which is minimized along the path that the object actually takes. Their simple approach led neatly to Newton's laws of motion for moving bodies (see page 8), and became known as 'the principle of least action'.

Soon it became clear that other theories in physics could also be derived by minimizing an action, including electromagnetism and general relativity (see pages 16 and 28). The principle was especially powerful when it came to combining theories. For example, calculating the behaviour of a body in the presence of both electromagnetic and gravitational forces simply involves summing the actions for both theories and then finding the path that minimizes this new, combined action.

NATURE IS THRIFTY IN ALL ITS ACTIONS.
Pierre-Louis Maupertuis

In the 1940s, American physicist Richard Feynman exploited the principle of least action to build his 'path integral' formulation of quantum field theory (see page 36). In this, the probability distribution for the state of a particle at some time in the future is given by summing the contributions over every possible path, weighted by the probability of that path being taken. The 20th-century American physicist Edwin Jaynes even proposed that there are deep connections between physics and information theory (see page 52).

Fermat's principle, and the principle of least action, remain some of the most powerful tools in physics and now lies at the heart of attempts to unify the forces of nature (see page 48) and to explain the very origins of our universe (see page 172).

The condensed idea
Light beams take the quickest path

02 Newton's laws

In 1687, Isaac Newton published a book that is widely regarded as seeding the birth of modern mathematical physics. At the heart of his revolution were three principles that encapsulated how objects behave under the influence of forces. These laws would dominate the physics of motion until the 20th century.

For hundreds of years, Newton's three laws of motion have offered the best description of how everyday objects move and interact – the branch of physics known as 'mechanics'. They were published, after much experimental and theoretical research, in 1687, as part of his book *Philosophiae Naturalis Principia Mathematica* (*Mathematical Principles of Natural Philosophy*), better known simply as the *Principia*. Prior to this time, mechanics had been dominated by the theories of the Greek philosopher Aristotle – which the burgeoning experimental science of the 17th century had by now proven to be deeply flawed. Not only was Newton's version of mechanics the first to be cast in terms of rigorous mathematical equations, but the answers the equations gave were spot on.

Newton's first law says that a body will continue in its present state of rest or uniform motion unless acted on by a force (an idea originally put forward by Italian mathematician and astronomer Galileo Galilei in 1632). In essence, an object that's stationary will remain stationary and an object that's already moving will keep moving at the same speed and in the same direction unless an external force is applied. You might wonder then, if you drop

TIMELINE

4TH CENTURY BCE	1021	1632
Greek philosopher Aristotle puts forward his ideas on the behaviour of moving objects	Persian philosopher Al-Biruni proposes the concept of acceleration as change in velocity	Italian astronomer Galileo publishes his ideas on the concept of inertia

this book, why it falls towards the ground, but that's because gravity exerts a constant downward force. Outside of gravitational fields, or in situations of effective 'zero gravity' such as Earth orbit, objects really do hang in space when released from rest, just as Newton predicted.

LINE OF FORCE

The second law quantifies how exactly the motion of an object changes when acted on by a force. Newton postulated that the object accelerates in the same direction as the force and at a rate that satisfies the mathematical equation force = mass x acceleration. This means that lighter objects will accelerate faster than heavier ones under the action of the same force: halve the mass of an object and it will accelerate twice as rapidly.

The resistance to motion of massive objects is called 'inertia'. You can think of it in terms of the first law. A body continues in its state of rest or uniform motion unless acted on by a force, and it's the inertia of the body – governed by its mass – that determines exactly how much that state of rest or uniform motion gets disturbed when the force acts.

Isaac Newton (1643–1727)

Isaac Newton was born at Woolsthorpe, a tiny hamlet in the English county of Lincolnshire. In 1661, he went up to Trinity College, Cambridge. He obtained his BA degree in 1665, before retreating to Woolsthorpe for two years to avoid that year's Great Plague. It was during this period of enforced isolation that he is said to have developed some of his most important ideas.

After returning to Cambridge in 1667, Newton was elected a fellow of Trinity College. Two years later, aged just 26, he became Lucasian Professor of Mathematics.

During his distinguished career Newton made seminal contributions not only to the physics of motion but also to gravity, optics, fluids, thermal physics and mathematics. He built the world's first reflecting telescope, and is even credited by some with inventing the cat flap. In 1703, he became president of the Royal Society, the world's oldest scientific society, and in 1705 he was knighted.

Newton never married, and he famously made many enemies, often over the priority of scientific discoveries. During later life, he worked as Warden and Master of the Royal Mint, where he prided himself in sending dozens of counterfeiters to the gallows. He died in his sleep on 31 March 1727. The death was registered as 'natural causes', though it's possible mercury poisoning incurred during his many alchemical experiments may have contributed.

1687
Isaac Newton publishes his three laws of motion in his book *Principia*

1750
Swiss mathematician Leonhard Euler extends Newton's laws to rigid bodies

1905
Einstein's special relativity marks the first major departure from Newton's laws

Newton's third law is all about how objects interact. It states that for every action there is an equal and opposite reaction. So when you sit on a chair, your weight pressing downwards under the action of gravity is balanced by an equal and opposite force pushing upwards from the chair. The source of this 'normal reaction', as it's known to physicists, is the network of chemical bonds between the atoms and molecules from which the chair is made. There's no guarantee, of course, that this chemical infrastructure will be up to the job – if you weigh too much, the chair will break and the normal reaction will vanish.

RECOIL REVEALED

It's Newton's third law that explains why a rifle kicks back against your shoulder when you fire it. Pulling the trigger releases the firing pin, igniting the powder in the cartridge. The expanding gases propel the bullet forwards but, thanks to Newton's third law, there's an equal and opposite force pushing the rifle back towards you. Incidentally, we can also invoke Newton's second law here – force = mass x acceleration – to explain why the bullet accelerates so much faster than the much heavier body of the gun.

Strictly speaking, Newton's laws only apply to objects whose masses are concentrated at a single point in space. They are a theoretical idealization

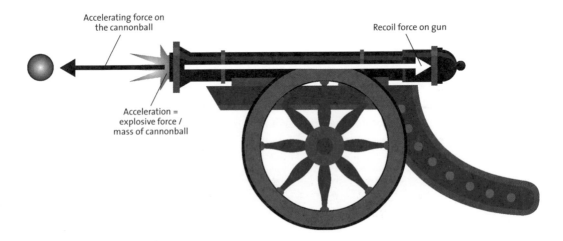

Accelerating force on
the cannonball

Recoil force on gun

Acceleration =
explosive force /
mass of cannonball

that makes calculations easier, but may not give the full picture. In or around 1750, the prolific Swiss mathematician Leonhard Euler (pronounced 'oiler') extended Newton's treatment to include rigid objects of non-zero size. He found that if you consider the mass of the object to be concentrated at its centre of gravity, then Newton's laws still apply. However, he also discovered additional laws governing how the object rotates – based on the turning forces, or 'torques', applied to the object and the precise distribution of its mass around the centre of gravity. The resulting Newton-Euler equations give an accurate description of objects in the real world.

NATURE AND NATURE'S LAWS LAY HID IN NIGHT: GOD SAID, "LET NEWTON BE!" – AND ALL WAS LIGHT.
Alexander Pope (1727)

MOVING ON

It would later emerge that even these laws don't quite give the full story in extreme cases. In 1905, Albert Einstein put forward his special theory of relativity (see page 24), in which the behaviour of objects travelling at close to light speed differs significantly from Newton's predictions. Later, Einstein's general theory of relativity (see page 28) led to further discrepancies in the presence of strong gravitational fields. Meanwhile, during the 1920s, it became clear that at the scale of subatomic particles of matter, the ordered deterministic view of Newtonian physics is replaced by the randomness of quantum mechanics (see page 32). Nevertheless, Newton's laws of motion remain an excellent approximation at the speeds and length scales, and in the ambient gravity, of our everyday world. In this regime, they have been verified by centuries of experimental study, and accurately describe the motion of everything from colliding billiard balls to planets orbiting the Sun.

The condensed idea
Moving bodies obey three mathematical rules

03 Newtonian gravity

In 1687, Isaac Newton published the first ever mathematical theory of gravity. It described everything from the motion of projectiles through the air to the orbits of the planets around the Sun – and, of course, the descent of falling apples. The theory has countless applications, ranging from air travel to satellite TV.

As well as his three laws of motion (see page 8), Isaac Newton's masterwork *Principia*, published in 1687, also contained the first rigorous treatment of the force of gravity. Cast with his trademark mathematical precision, Newton's law of universal gravitation explained phenomena such as balls rolling downhill and the orbits of the planets. The law states that the gravitational force between two massive bodies is proportional to their masses multiplied together, divided by the square of the distance between them. Double either of the masses and the force doubles, too. However, double the distance between them and the force diminishes by a factor of four. It's a relatively simple mathematical relationship – one that Newton had arrived at after studying the behaviour of falling objects and astronomical data on the motion of the planets.

FALLING APPLES

The law says that an object released from rest on Earth, such as an apple dropping from a tree, accelerates towards the ground at a rate determined purely by the planet's mass and size. At the surface of the Earth, the rate of acceleration due to gravity is 9.8 metres per second (32 ft per second) every second. That is, for every second of free fall, the downward speed increases by 9.8 metres per second – the mass of the object being dropped is irrelevant. Objects thrown vertically

TIMELINE

1609-1619	1666	1687
German mathematician Johannes Kepler publishes his three laws of planetary motion	Robert Hooke presents his early ideas on gravity to the Royal Society	Isaac Newton introduces his full theory of gravity in his book Principia

into the air will initially lose speed at the same rate before falling back to the ground. And those with a horizontal component to their velocity arc through the air on a curved trajectory, called a parabola, which brings them back to Earth at a point some way away from where they started.

Applying Newtonian gravity to projectiles on Earth is only an approximation in practice, because our atmosphere introduces air resistance – a drag force that slows moving objects down, significantly so at high speeds. Air resistance caps the speed of a falling object at a so-called 'terminal velocity' dependent on its aerodynamic properties. For example, a skydiver free-falling head first towards the ground has a terminal velocity of up to 530 kilometres per hour (330mph). Opening their parachute slows this to 28 kilometres per hour (17mph). On the Moon, however, where the atmosphere is negligible, Newton's predictions are bang on – as US astronaut Alan Shepard demonstrated, having smuggled a 6-iron aboard the Apollo 14 lunar lander mission in 1971. As Shepard put it, the golf ball he struck went for 'miles and miles and miles'.

Newton's law of gravity

Mathematically, Newton's law of universal gravitation says that if two bodies, with masses m_1 and m_2 are separated by a distance r, then each body experiences a force of attraction g towards the other, given by the formula:

$$g = G\,\frac{m_1 m_2}{r^2}$$

Where G is the gravitational constant, 6.67×10^{-11} (6.67/100,000,000,000). The force g is measured in Newtons, where 1 Newton will cause a mass of 1 kilogram (2.2 lb) to increase in speed by 1 metre (3.3 ft) per second every second.

INTO ORBIT

Keep increasing the speed of a projectile launched from Earth and eventually it won't come back down. Instead, it will reach orbit – circling the planet indefinitely. Add more speed and the orbit grows wider and wider until the Earth's gravity can no longer hold on to it and the projectile flies off into deep space. Newton was particularly proud of this 'universality' of his theory – the fact it applies equally well at the Earth's surface as it does millions of miles

1798
British physicist Henry Cavendish carries out first lab test of Newtonian gravity

1916
Albert Einstein's general relativity replaces Newton's theory in extreme cases

1945
Using Newtonian gravity, Arthur C. Clarke lays the foundations for satellite communication

The cannonball experiment

Isaac Newton devised a thought experiment to demonstrate that his new law of gravity was just as good at describing the planets orbiting the Sun as it was apples falling from trees He imagined a cannon fired horizontally from a high mountaintop. The trajectory of the cannonball is determined by its speed and the pull of Earth's gravity.

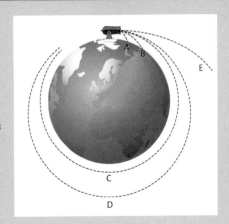

For relatively slow speeds, the ball arcs downwards as gravity pulls it to the planet's surface (path A on the diagram). Increase the speed of the ball and the point where it lands gets progressively further away from the mountain (e.g., path B). Eventually, the speed of the ball is high enough that the planet's spherical surface falls away at the same rate the ball is falling towards it, and the ball never reaches the ground. Instead, it circles in orbit around the Earth (path C). Higher speeds still stretch the orbit into a wide ellipse (path D), until finally the Earth's gravity can no longer hold on to the cannonball and it flies off into space (path E). The speed required to do this, called the 'escape velocity', is determined purely by the planet's gravity and is independent of the mass of the cannonball: in Earth's case it is 11.2 kilometres per second (7 miles per second).

away on the fringes of the Solar System. He was able to demonstrate this by using the theory to mathematically derive Kepler's laws of planetary motion (see page 167).

Not everyone agreed that Newton's work was a triumph, however. The British natural philosopher Robert Hooke accused Newton of plagiarism, claiming that he had already come up with the idea of modelling gravity as an 'inverse square law' (the $/r^2$ bit of Newton's formula – see box on page 13). It's true that Hooke did indeed present ideas on gravity to the Royal Society as early as 1666, but modern historians believe the idea of gravity following an inverse square law had been well established for some time, so neither Hooke nor Newton deserves full credit. What is beyond dispute is that Newton was the first to go the extra mile by incorporating the inverse square law into a full working theory, and demonstrating its correctness.

THE BIG G

The first laboratory test of Newtonian gravity wasn't carried out until 1798 – 70 years after Newton's death – by the British physicist Henry Cavendish. Cavendish used a piece of apparatus called a 'torsion balance', which measures twisting forces on a wire, to measure the tiny gravitational attraction between lead weights. The results of the experiment verified the theory and led to the first accurate determination of the constant of proportionality G in Newton's equation – currently estimated at 6.67×10^{-11} m³/kg/s².

I CAN CALCULATE THE MOTION OF HEAVENLY BODIES, BUT NOT THE MADNESS OF PEOPLE.
Sir Isaac Newton

Newtonian gravity would prevail for another 120 years, until Albert Einstein published his general theory of relativity, a new approach involving curvature of space and time (see page 28). Einstein resolved various anomalies that were emerging with Newton's theory, such as failure to explain the strange orbit of the planet Mercury and the bending of light rays around the Sun. Nevertheless, the predictions of general relativity only differ significantly from Newton's model in strong gravitational fields, and Newton's simpler theory can still be used today in applications ranging from weather prediction to computing the trajectories of spacecraft.

In particular, the British science fiction writer Arthur C. Clarke realized in the 1940s that the Newtonian laws mean there is a particular altitude above the Earth at which a satellite will orbit at the same rate that the planet is rotating – completing one orbit every 24 hours. Viewed from the ground, the satellite will appear to hang in the sky. These 'Clarke orbits', 35,786 kilometres (22,236 miles) above sea level, now form the basis of satellite communications, used daily to beam information and entertainment around the globe.

The condensed idea
What goes up must come down. Maybe

04 Electromagnetism

One of the greatest technological tools of the modern age is mobile communication via radio signals. Yet this is only possible thanks to a string of discoveries in the 19th century, culminating in James Clerk Maxwell's realization that electricity and magnetism are just different aspects of the same thing.

One of the earliest clues to the connection between electricity and magnetism emerged from the experiments of Danish physicist Hans Christian Ørsted. In 1820, he noted that passing an electric current through a wire moved the needle of a nearby magnetic compass. By moving the compass, Ørsted found that the lines of magnetic force trace out circles around the wire. Later that same year, the French physicist André-Marie Ampère learned of Ørsted's discovery and set about formulating a piece of theory to explain it. He saw that the magnetic fields around two parallel wires make the wires attract each other if the currents are going the same way, and repel each other when they're moving in opposite directions. He was able to capture this behaviour in a mathematical relationship, now known as Ampère's law, which gives the magnetic field around each wire in terms of the current producing it – and allows calculation of the force between the two.

CURRENT AFFAIRS

In 1831, British physicist Michael Faraday demonstrated the converse effect. He connected a battery to a coil of insulated wire wrapped around one side of an iron ring. Around the other side of the ring, he wound a second coil of insulated wire, the ends of which were connected to a galvanometer (an

TIMELINE

1820	1820	1831
Ørsted discovers that electric currents cause magnetic fields	Ampère develops a theoretical explanation for Ørsted's observations	Faraday demonstrates how magnetic fields can create electric current

instrument for measuring electric current). When he switched on the left-hand coil, Faraday noticed a brief current in the galvanometer on the right. The left-hand coil had created a magnetic field in the iron, which had in turn induced a current in the right-hand coil. The same thing happened when an ordinary magnet was moved near the right-hand coil. Faraday concluded that a time-varying magnetic field produces an electric current, an effect now known as induction. He utilized the principle to build the first electric generator – the dynamo.

> I HAPPEN TO HAVE DISCOVERED A DIRECT RELATION BETWEEN MAGNETISM AND LIGHT, ALSO ELECTRICITY AND LIGHT, AND THE FIELD IT OPENS IS SO LARGE AND I THINK RICH.
>
> Michael Faraday

The breakthrough that was to completely reveal the complex interplay between electricity and magnetism was made in 1861 by Scottish physicist James Clerk Maxwell. Working at King's College, London, Maxwell distilled the findings of Faraday, Ampère, Ørsted and others into four coupled mathematical equations. Take any configuration of electric charge and/or electric current, and the equations tell you exactly how the resulting electric and magnetic fields behave. They would become known as Maxwell's equations.

The first two equations summed up earlier work by the German physicist Carl Friedrich Gauss. He had shown that the electric field around a static electric charge increases in direct proportion to the size of the charge – this was Maxwell's equation 1. Equation 2 essentially said that the net magnetic field around a point is zero. Equation 1 meant that lines of electric field around an electric charge are all directed radially outwards – so that there is a net electric field leaving the charge. However, equation 2 stated that the same is not true for magnetism. It was another way of saying that whereas isolated electric charges ('monopoles') can exist freely, magnetic poles must always come in pairs ('dipoles'), so that the field leaves at one pole and enters at the other. This marries up with our experience: magnets always come with a north and south pole.

1835
Gauss explains electric and magnetic fields around static electric charges

1861
Maxwell formulates his four equations unifying electricity and magnetism

1864
Maxwell uses his new theory to predict the existence of electromagnetic waves

Maxwell's third equation is a restatement of Faraday's discovery that the electric field around a point (or equivalently the current flowing through a circuit) is determined by the rate of change of the magnetic field with time. Meanwhile, the fourth and final equation was an extension of Ampère's law – which simply said that the magnetic field around a point is given by the current. In keeping with his third equation, Maxwell amended Ampère's law to include a contribution to the magnetic field from any time-varying electric fields present.

Maxwell's realization that not only do changing electric fields make magnetic fields, but also changing magnetic fields make electric fields, soon led to another momentous breakthrough. He showed that it's possible for a pair of time-varying electric and magnetic fields to exist together, in empty space, in the absence of any electric charges or currents. The magnitude of each field rises and falls in a steady rhythm – with the oscillations of one driving the other via Maxwell's third and fourth equations. Maxwell had discovered electromagnetic waves, and when he worked out the speed of these waves through space – which, from his equations, turned out to be just a simple combination of the electric and magnetic properties of empty space – the number he got was extremely close to existing measurements of the speed of light (currently 299,792,458 metres per second (186,282 miles per second). His conclusion, which he published in 1864, was simple: light itself is an electromagnetic wave.

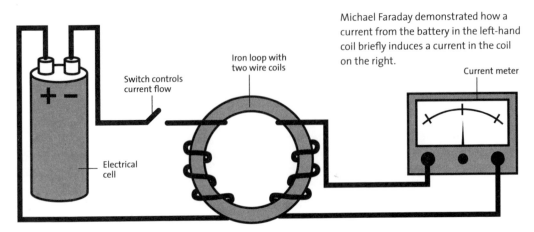

Michael Faraday demonstrated how a current from the battery in the left-hand coil briefly induces a current in the coil on the right.

Iron loop with two wire coils

Current meter

Switch controls current flow

Electrical cell

WAVE POWER

Visible light has a wavelength of between 380 and 760 nanometres (millionths of a millimetre). But Maxwell's theory predicted the existence of waves at all wavelengths. Some were already known, such as infrared and ultraviolet, but others awaited discovery. In 1888, German experimental physicist Heinrich Hertz used apparatus inspired by Maxwell's theory to generate and detect electromagnetic waves with wavelengths between a millimetre to many kilometres. They became known as radio waves and are central to most of the wireless communication that we enjoy today.

Like Newtonian mechanics, Maxwell's classical theory of electromagnetism would, in the early to mid-20th century, be surpassed by a new treatment (see page 36). But Maxwell's equations were a revolution in their own right. They cemented with physicists not only the important concept of fields, but also that of 'unification' – the idea of consolidating all the fundamental forces of nature into one overarching theory of everything, a pursuit that now lies at the very heart of modern physics research (see page 48).

James Clerk Maxwell (1831–79)

Maxwell was born in Edinburgh in 1831. He was educated by his mother until the age of eight, by which time he had already demonstrated prodigious intellect. When his mother died, he was sent to the Edinburgh Academy. He struggled to fit in there at first, but by the age of 13 was winning awards. He published his first scientific research paper aged 14. It concerned drawing geometric curves using string, and was presented on his behalf to the Royal Society of Edinburgh.

In 1847, Maxwell went to study at the University of Edinburgh. He published two further papers there, before moving to Cambridge in 1850, where he remained until 1856. After four years running a department at Marischal College, Aberdeen, he took up a professorship at King's College, London, where he carried out his seminal work on electromagnetism.

Maxwell was married to Katherine Mary Dewar in 1858, but they had no children. He died of abdominal cancer in 1879, at the age of just 48, and is buried at Parton Kirk in Galloway, Scotland.

The condensed idea
Electricity and magnetism are different sides of the same coin

05 Thermodynamics

Driven by the need to make steam engines more efficient in the industrial revolution, scientists developed our understanding of the complex interplay between heat, energy and motion. Today, the science of thermodynamics is applied in everything from aircraft design to speculations about the fate of the universe.

Temperature and movement are intimately connected, as you'll be aware if you've ever noticed how a bicycle pump gets warm as you use it, or a spray deodorant feels cold as its expanding vapour hits your skin. Thermodynamics is the branch of physics that governs these processes and a raft of others. It determines how heat (the energy locked away in a physical system because of its temperature) gets converted into 'work' (physicists' speak for useful mechanical movement). For example, when you compress the bicycle pump, you're applying mechanical work to the air inside, which gets converted into heat and raises its temperature. Meanwhile, the spray deodorant is doing the opposite, converting heat into mechanical work, and cooling down as the gas expands.

Thermodynamics began in the late 17th century when British physicists Robert Hooke and Robert Boyle, and German inventor Otto von Guericke, independently built mechanical air pumps. Studying the behaviour of these contraptions led Boyle and Hooke to realize that a simple law governed the behaviour of a gas – namely, that its pressure is inversely proportional to its volume: reduce one, and all other things being equal, the other goes up.

TIMELINE

1698	1738	1824
Thomas Savery patents an early steam engine design	Daniel Bernoulli publishes the foundations of kinetic theory	Sadi Carnot publishes his work on the efficiency of steam engines

AGE OF STEAM

In 1697, English inventor Thomas Savery built on these concepts to create the first working steam engine – a device to convert the heat in steam into motion of a piston rod. In 1712, Thomas Newcomen refined these ideas further to build an engine that was used commercially across Britain. In the late 17th century Scottish engineer James Watt made further modifications – but even Watt's engine could only convert around 3 per cent of the energy in steam into useful work. It was a French military engineer named Sadi Carnot who figured out why. In 1824, he realized that steam engines require a difference in temperature in order to operate – the piston moves because hot gas on one side is expanding into cold gas on the other. Increase the temperature difference, and the efficiency of the engine goes up, too. Thanks to this discovery, steam engine efficiency leapt to 30 per cent by the mid-19th century.

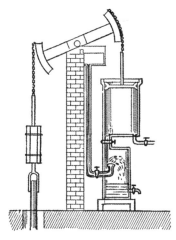

But what about the energy that's not converted into work? In 1850, German physicist and mathematician Rudolf Clausius came up with the concept of 'entropy' to describe this essentially useless residual heat. Entropy can be thought of as the 'degree of disorder' of a heat engine. Low entropy means the engine is in a highly ordered state, with well-defined hot and cold sides (corresponding to an engine that is highly efficient). On the other hand, a high-entropy engine has a largely uniform temperature: very little heat flows, and very little useful work can be extracted.

In Watt's engine design, steam from the boiler pushes the piston upwards. At the top of the stroke, the condenser valve opens and the hot steams floods into the condenser where it is rapidly cooled with cold water. Atmospheric pressure then drives the return stroke, pushing the piston back to its starting position.

STATE OF DISORDER

Clausius further deduced that entropy must always increase. Place a hot and a cold object side by side (a state of low entropy) and heat will flow

1850

Clausius gives the first statement of the first and second laws of thermodynamics

1906-12

Walther Nernst presents a formulation of the third law of thermodynamics

1935

The term 'zeroth law of thermodynamics' is used for the first time from the hot body to the cooler one until both are at the same temperature (high entropy). He enshrined this observation in what is now known as the second law of thermodynamics: entropy increases. Another good example is a box with a partition down the middle, either side of which is a different type of gas. This is a highly ordered, low-entropy state.

> **NOTHING IN LIFE IS CERTAIN EXCEPT DEATH, TAXES AND THE SECOND LAW OF THERMODYNAMICS.**
> Seth Lloyd

Take away the partition and slowly the gases begin to diffuse together. Eventually the whole box will be filled with a uniform mix of both gases – this is a high-entropy, disordered state. The converse, a mixture of gases spontaneously separating, is so unlikely that it will never happen.

However, it's always possible to violate the second law locally. For example, let's say the inside of your car has the same temperature as the outside – a high-entropy state. Switching on the air conditioning creates a big temperature difference and an apparent reduction in entropy. But on a larger scale, the car's engine has to do extra work to power the air con, and because the engine isn't 100 per cent efficient, the entropy rise this creates is always greater than the localized drop inside the car. Thus the entropy of the whole universe increases, and the second law holds.

LAW ENFORCEMENT

There are in total four laws of thermodynamics. The first says that the increase in a body's internal heat energy is just the heat entering the body minus any work that it does. The third law says that zero temperature also corresponds to zero entropy. The 'zeroth' law, recognized after the other three, but considered more fundamental, says that if body 'A' is in thermodynamic equilibrium with body 'B' (so that no heat flows between the two) and body 'A' is in equilibrium with body 'C', then 'B' must also be in equilibrium with 'C'.

Although pioneered for the development of mechanical systems, thermodynamics soon found other applications – describing phenomena such as heat transfer during chemical reactions and the behaviour of weather systems. Entropy has also become a powerful concept in statistics and information theory (see page 52).

Small world

Classical thermodynamics describes the large-scale, 'macroscopic' behaviour of bodies, in terms of bulk quantities like temperature and pressure. In his 1738 book Hydrodynamica, Swiss mathematician Daniel Bernoulli began to apply thermodynamics to much smaller units – atoms and molecules. He outlined what has since become known as kinetic theory – the idea that the temperature of a gas is due to the individual motions of its constituent particles. The physical sensation of heat is actually caused by these particles drumming against your skin.

Later it was realized that not all particles in a gas move at the same speed. Scottish physicist James Clerk Maxwell and Austrian Ludwig Boltzmann came up with a statistical distribution giving the number moving at any particular speed, for a given temperature. It was the birth of 'statistical mechanics' – the application of probability and statistics to the microphysics of matter in order to make predictions about its macrophysics.

In the 20th century, statistical mechanics took another quantum leap forward, quite literally – incorporating the newly discovered laws of quantum mechanics, which govern the behaviour of subatomic particles (see page 32). Probability operates very differently in the quantum world. And this, along with other new and exclusively quantum behaviours, drastically alters the resulting thermodynamics.

Astrophysicist Stephen Hawking and colleagues have found close parallels with the physics of black holes (see page 188), describing them with four mathematical laws that mirror the four laws of thermodynamics. Meanwhile, the inexorable increase of entropy dictated by the second law of thermodynamics on a cosmic scale (see page 184) has been adopted as one possible scenario for the death of our universe.

The condensed idea
Heat obeys a strict set of physical laws

06 Special relativity

In 1905, an unknown German scientist called Albert Einstein changed our view of the world forever by rewriting the centuries-old laws governing the behaviour of moving objects. According to Einstein's theory of special relativity, motion at close to the speed of light can distort reality and even slow the passage of time.

Einstein's relativity wasn't an entirely new idea. As early as the 17th century, Italian scientist Galileo Galilei had been the first to realize that the perceived motion of an object depends crucially on the motion of the observer. If you're driving a car at 50 kilometres per hour (30 mph), then traffic coming the other way at the same speed will appear, from your frame of reference, to be heading towards you at 100 kilometres per hour (60 mph). Similarly, if you pull alongside another car travelling in the same direction at the same speed then, again relative to your own motion, it will appear stationary. Galileo came up with mathematical rules to calculate the speed of one object relative to the other.

At the end of the 19th century, however, a young Albert Einstein began to wonder what would happen if one of these moving objects was replaced with a light beam. What if you could travel very fast alongside the light beam to see how it looks when stationary?

Aside from lacking a vehicle that could propel him at the necessary 300,000 kilometres per second (186,000 miles per second), Einstein realized that there was another fundamental problem with this idea.

TIMELINE

1632	1818	1864
Italian physicist Galileo proposes his early version of relativity	French scientist Augustin-Jean Fresnel develops the wave theory of light	James Clerk Maxwell proves the constancy of the speed of light

According to James Clerk Maxwell's theory of electromagnetism (see page 16), the speed of light is a universal constant of nature. That is, it should be the same for all observers – whether stationary or on a speeding rocket. There had to be something wrong with Galilean relativity, but since it works fine in our everyday experience Einstein reasoned that the problems must begin close to light speed itself.

INTO THE ETHER

Ever since Augustin-Jean Fresnel's development of the wave theory of light in the early 19th century, scientists had assumed that light waves travelled through some kind of medium – just like waves in water or on a string. But this medium, known as the 'luminiferous ether', had proved hard to detect. In 1887, two American physicists – Albert Michelson and Edward Morley – set up an experiment to try and measure the Earth's motion through it. Their failure prompted a flurry of research to try and explain why. In particular, Dutch physicist Hendrik Lorentz realized that the result could remain consistent with the existence of an ether if moving objects contracted slightly in their direction of motion. He even came up with mathematical formulae to support the idea.

Albert Einstein (1879–1955)

Albert Einstein was born in the German city of Ulm on 14 March 1879. After emigrating to Switzerland in 1896 to avoid military service, he studied mathematics and physics at Zurich Polytechnic before taking a position as a clerk in the patent office in Bern. It was in his spare time here that he carried out much of his work on special relativity. The theory was published in 1905, and followed up with general relativity in 1915.

In 1921, Einstein was awared a Nobel Prize – though this was for his work on the photoelectric effect (the generation of electricity by light) rather than relativity. Much of his later professional life was dedicated to a fruitless search for a unified theory of the forces of nature (see page 48).

Einstein married twice, and is rumoured to have had many affairs. He had three (known) children. Being Jewish, he moved to the USA during the 1930s to escape the rise of fascism in Europe. He ultimately settled at Princeton University, where he died on 18 April 1955, aged 76.

1887
Michelson and Morley fail to detect a medium through which light waves travel

1895
Dutch physicist Hendrik Lorenz's length contraction formulae are published

1905
Einstein publishes his completed special theory of relativity

We now know that there is no such thing as ether: as a wave of electromagnetic energy, light can move through a vacuum without the need for a medium. The failure of Michelson and Morley was inevitable, but Lorentz's work didn't go to waste: Einstein realized that his formulae were exactly what he needed to square the discrepancy between Galilean relativity and the constancy of the speed of light demanded by Maxwell's electromagnetism. The formulae involved scaling certain quantities from Galilean relativity by a factor equal to $\sqrt{1-v^2/c^2}$, where v is the speed of the moving object and c is the speed of light. In this framework, at low speeds (small v), the predictions of Galileo's theory hold, but as v approaches c the results become very different.

> THERE WAS A YOUNG LADY NAMED BRIGHT, WHOSE SPEED WAS FAR FASTER THAN LIGHT; SHE SET OUT ONE DAY IN A RELATIVE WAY, AND RETURNED ON THE PREVIOUS NIGHT.
>
> A. H. Reginald Buller

SHORT MEASURE

As Lorentz had originally shown, the new theory demanded that fast-moving objects contract in their direction of motion. For instance, the length of a spacecraft travelling at 86 per cent light speed would be halved. But there was even stranger to come. Special relativity placed space and time on an equal footing, just different dimensions in a four-dimensional fabric called 'spacetime'. For this reason, Lorentz's formulae mean that time also gets distorted by the $\sqrt{1-v^2/c^2}$ factor. On board the same spacecraft, travelling at 86 per cent of light speed, clocks tick only half as often as they do when stationary. If the spacecraft left Earth and travelled at this speed for a year (as measured by clocks on board), the astronauts would return to find that twice as much time – two years – had elapsed back home. This effect is called 'time dilation'.

The weird predictions of special relativity are confirmed regularly in particle physics experiments (see page 40), where subatomic particles are accelerated to near-light speed and smashed together, creating showers of exotic new particles. Many of these new particles are unstable and decay on predictable timescales. Predictable, that is, in the particles' own reference frames: at the immense speeds involved, the decay times are seen to be stretched out, exactly in accordance with time dilation.

The unity of spacetime was to have one other profound repercussion. In the three dimensions of ordinary space, moving objects have a property called kinetic energy – energy that they possess by virtue of their motion. When Einstein extended this notion into four dimensions, he found that objects also possess energy as a result of their inexorable motion through time. Even a stationary object will have this so-called rest mass energy, given by the now-famous formula $E = mc^2$, energy equals mass multiplied by the speed of light squared.

Whenever mass is lost, energy is released and Einstein's formula reveals exactly how much. For example, if a piece of wood is burned, then taking the mass difference (the starting mass of the wood minus the mass of the residual ash and smoke) and multiplying it by c^2 gives the energy released.

Faster than light?

Strange things happen to objects moving near the speed of light. But what happens when you go faster than light? Einstein found that as an object accelerates, its effective mass increases. This adds to the object's inertia or resistance to motion, making it harder to accelerate any further. At the speed of light the effective mass is infinite, meaning it's impossible for any object with mass to reach light speed: special relativity enforces a cosmic speed limit.

What's more, at light speed, time dilation means that time literally stands still. If you could go faster than light, you would actually see time start to run backwards, but special relativity also seems to rule out time travel.

LET'S SPLIT

In the late 1930s, scientists noticed that splitting an atomic nucleus of the heavy element uranium produced two new nuclei whose masses didn't quite add up to the mass of the uranium. If all the nuclei in a lump of uranium could be split this way, then the implication was that a truly massive amount of energy would be released. Relativity had thus enabled the discovery of nuclear power (see page 44). Today, the theory is a guiding principle in fundamental physics research on scales ranging from the realm of subatomic particles to the universe at large.

The condensed idea
Your view of reality depends upon your state of motion

07 General relativity

After ten years of intense effort, Albert Einstein incorporated the force of gravity into his theory of relativity. The new model of general relativity worked by curving the flat space and time of the earlier special theory. It explained anomalies in Newton's law of gravity – and gave the world satellite navigation.

Einstein's special theory of relativity (see page 24) was an extraordinary triumph of the imagination. But it was, by design, incomplete, applying only to the special case of objects moving at constant speed. Following the theory's publication in 1905, Einstein set about extending it to the more general case of objects undergoing arbitrary acceleration, for example those falling under gravity.

The best theory of gravity, Isaac Newton's universal law of gravitation, had held sway since its publication in 1687 (see page 12). But Newton's model was blatantly at odds with the principles of relativity, holding that the gravitational force propagates across space instantaneously while relativity says that nothing can travel faster than light.

G-FORCE

Earlier experiments by Italian scientist Galileo led Einstein to suspect that the action of gravity could be summed up purely as a rate of acceleration. (Galileo had dropped balls of differing weights from the top of the Leaning Tower of Pisa – proving that they fell at the same rate.) Einstein concocted a thought experiment in which balls are being fired across the inside of

TIMELINE

1687	1854	1905
Sir Isaac Newton publishes his law of gravity as part of his book *Principia*	German mathematician Bernhard Riemann completes his thesis on differential geometry	Einstein publishes special relativity, describing objects moving close to light speed

a sealed chamber. He realized that if you were watching this, you'd have no way of telling if the arcs the balls made through the air were caused by gravity, or whether the entire box was in zero-gravity but accelerating upwards under its own power: the observable effects of both are identical. Einstein once described this realization as 'the happiest thought of my life'.

A further thought experiment led him to see the connection between acceleration and curvature. Einstein imagined a rapidly spinning disc: keeping an object moving on a circular path requires an acceleration directed towards the centre of the circle. Because of the 'length contraction' effect of special relativity (see page 26), the outer circumference of the disc will shrink, and the only way this can happen while leaving the radius of the disc unchanged is if the disc curves into a bowl shape.

> NO MATTER HOW HARD YOU TRY TO TEACH YOUR CAT GENERAL RELATIVITY, YOU'RE GOING TO FAIL.
> Brian Greene

IN SEARCH OF THE FIELD EQUATION

Einstein was thus convinced that he could explain gravity by adding curvature to the flat four-dimensional 'spacetime' of special relativity. The big question was to figure out exactly how the curvature is determined by contents of space. The search for an answer to this question occupied him until 1915.

The object of Einstein's quest was a relationship known as the 'field equation', essentially equating measures of curvature on one side with 'source terms' on the other. In Newtonian gravity, the source terms simply involve mass. Special relativity says that mass and energy are equivalent (from the famous equation $E = mc^2$) so you might expect both mass and energy to feature in a relativistic theory of gravity. In fact, Einstein found that mass, energy, pressure and momentum all contribute.

1915
Einstein's general theory of relativity builds gravity into the special theory

1919
Arthur Eddington measures light bending by the Sun, confirming general relativity

2016
The international LIGO collaboration announces the discovery of gravitational waves

With the field equation in place, physicists can work out exactly how space and time are curved by the distribution of matter. Then the trajectories of moving objects can be calculated – just like the path of a marble rolling across the dips and peaks of a curved rubber sheet.

Confirmation that Einstein had indeed found the correct field equation was soon to follow. According to general relativity, gravity bends the paths not only of solid objects, but also of light rays. Einstein predicted that starlight passing close to the Sun should be bent by 0.0005 of a degree. The trouble was seeing this tiny effect through the Sun's fierce glare. British astronomer Arthur Eddington had the solution – to measure light bending during a total solar eclipse, when the Sun's light is obscured by the Moon. In 1919, Eddington travelled to the West African island of Príncipe, where his observations of starlight during an eclipse were in good agreement with general relativity.

Calculating curvature

In developing his curved-space theory of gravity, Einstein needed to find mathematical tools capable of quantifying the curvature of four-dimensional space and time.

Luckily for him, during the latter half of the 19th century, the German mathematician Bernhard Riemann had put forward his seminal ideas on 'differential geometry' – a way of using algebra to extend the basic principles of geometry in two-dimensional flat space to higher dimensional spaces of arbitrary curvature.

In the four dimensions of space and time, Riemann's formalism boiled down to assigning ten numbers to every point in space, bundled up into a mathematical object called a 'tensor'. The field equation of general relativity then determined the tensor's components from the material content of space.

Today, the bending of light by gravity is observed on much larger scales. When a chance alignment causes the light from a distant galaxy to be focused by the gravity of an intervening galaxy or cluster of galaxies, the result is called a gravitational lens. The distant galaxy is magnified by the lens and multiple images of it are often seen. The first real gravitational lens was discovered in 1979, in the constellation of Ursa Major.

MISBEHAVING MERCURY
General relativity also cleared up an old mystery. In 1859, French astronomer Urbain Le Verrier had noticed that the orbit of the planet Mercury wasn't behaving as it should. The oval path of the planet's orbit was itself rotating around the Sun, tracing out a rosette-like pattern over time. Newtonian gravity was at a loss to explain why, but when the planet's

orbit was recalculated using general relativity this 'perihelion precession' dropped neatly out of the mathematics.

Perhaps the greatest evidence for the theory, however, is observed daily by GPS users. General relativity is essential to the operation of satellite navigation systems. That's because signals beamed from orbit down to the Earth's surface gain a small amount of energy, as they 'fall' in our planet's gravitational field. A GPS unit uses Einstein's mathematics to correct the satellite time signals for this effect. Without it, the location would be out by miles.

Until very recently, there was a missing piece in the puzzle. General relativity says that strong, time-varying sources of gravity, such as black holes orbiting closely around one another, must give out 'gravitational waves' – ripples in space itself. A passing gravitational wave causes a brief but measurable distortion in the distance between two points. The distortion is so small it has taken scientists a full century to detect the waves (see box, right). The discovery has opened a whole new window on the universe, which promises new insights into the physics of black holes (see page 188) and the Big Bang (see page 172).

The hunt for gravitational waves

On 11 February 2016, an international collaboration of astronomers announced the historic detection of gravitational waves. The Laser Interferometric Gravitational Wave Observatory (LIGO) project saw a blast of gravitational waves given off as two black holes in the southern hemisphere of the sky merged into one.

The signal was picked up simultaneously by two giant laser interferometer detectors – one in Livingstone, Louisiana, and the other in Hanford, Washington State. Each detector consists of a two-armed L-shaped structure, with each arm measuring several kilometres in length. Laser light is beamed down each arm and reflected by mirrors back to the apex of the 'L', where the beams are combined to form an interference pattern. The passage of a gravitational wave causes tiny changes in the relative lengths of the arms, which shift the interference pattern. Each LIGO detector saw exactly the same shift, ruling out random noise.

The condensed idea
Gravity is caused by the curvature of space and time

08 Quantum mechanics

In the 19th century, scientists got their first clues to suggest that something was wrong with the laws of mechanics – the branch of science dealing with how objects move under the action of forces. It was becoming clear that a new theory was needed down at the minuscule scale of atoms.

It all started with a revolution in the understanding of light. Physicists had long argued over whether light was made up of waves or particles. In 1803, the British physicist Thomas Young seemed to have proved beyond all doubt that light beams are indeed waves, demonstrating that two beams can form ripple-like interference patterns, just like colliding water waves. But then, in 1905, Albert Einstein rekindled the debate when he presented a compelling explanation of the photoelectric effect – the way that certain metals can be made to produce electricity in response to light.

PLANCK'S QUANTA

Einstein was inspired by the work of his German colleague, the physicist Max Planck. A few years earlier, Planck had successfully explained the connection between the temperature of a hot object and the frequency of the light it gives off (explaining why a heated poker changes from orange to white as it gets hotter). This problem had long baffled physicists, and Planck's insight was to suppose that light was emitted in discrete chunks, with the energy of each chunk given by its frequency multiplied by 6.63×10^{-34}, a number known today as Planck's constant.

TIMELINE

1803	1900	1905
Thomas Young demonstrates that light behaves like a wave	Max Planck explains thermal radiation with a particle theory of light	Albert Einstein uses Planck's theory to explain the photoelectric effect

This assumption led straight to the right answer, but Planck himself viewed this as no more than a quirk in the behaviour of light waves and their interaction with matter. Einstein, however, took the description literally, interpreting the chunks as actual particles, or 'quanta', of light. This was the model he brought to bear on the photoelectric effect. The nub of the problem was to explain why only light above a certain frequency is seen to create an electric current.

ANYONE WHO IS NOT SHOCKED BY QUANTUM THEORY HAS NOT UNDERSTOOD IT.
Niels Bohr

In Einstein's view this was now straightforward: the light quanta collide with electrons in the metal like billiard balls, but only quanta with sufficient energy can dislodge them to create a current – and because of Planck's formula, this can only happen when the light's frequency is high enough. Planck hated the idea, but quanta of light – 'photons', as they were subsequently named – were discovered experimentally in 1923 by the American physicist Arthur Compton.

WAVE-PARTICLE DUALITY

Einstein and Compton had proved that light was made of particles, while Young, a century earlier, had shown conclusively that it's made of waves. The only way forward was if they were both right. This seemingly bizarre notion was cemented by French physicist Louis de Broglie. In 1924, he came up with a formula linking the wavelength of light to the physical momentum of its photons. He soon realized that the same formula, turned on its head, could be used to infer a wavelength for objects previously regarded as solid particles, including the protons and electrons inside atoms (see page 68).

De Broglie's work became known as the theory of wave-particle duality, and was confirmed in 1928 by researchers at Bell Labs, New Jersey. They fired a beam of electrons at a crystalline grid. They had calculated the de Broglie wavelength of the electrons, finding it to be similar to the size of the gaps in the grid. When waves pass through a grating spaced at intervals comparable to their wavelength they undergo a phenomenon called diffraction, where

1923
Photons are discovered experimentally by American physicist Arthur Compton

1924
Louis de Broglie shows that solid particles also behave like waves

1926
Schrodinger encapsulates de Broglie's idea with a wave equation for particles

Schrödinger's cat

The view of quantum mechanics that emerged in the 1920s argued that subatomic particles are described by a wave function, which gives the probability of finding the particle at any particular point in space. When a measurement is made, the wave function 'collapses' and the particle is seen with a definite position. This became known as the 'Copenhagen interpretation' because many of the details were thrashed out at a conference in Copenhagen in 1927.

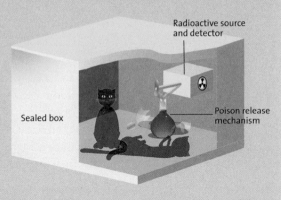

Radioactive source and detector

Sealed box

Poison release mechanism

Erwin Schrödinger, however, felt very uncomfortable with a view of reality in which the state of a particle depends on whether it's been measured or not. To highlight the absurdity, he concocted a famous thought experiment. He imagined a cat locked in a box with a phial of deadly poison whose release is controlled by a subatomic particle detector. If a particle passes through the detector, the poison is released and the cat dies, but otherwise it lives. Because the detection of a particle is a quantum process governed by probability then, according to the Copenhagen interpretation, the cat must be simultaneously alive and dead until the box is opened and a measurement made.

Nowadays, physicists shun collapse of the wave function in favour of a picture known as 'decoherence', where the transition from wave to particle behaviour is due to the interaction of a fragile quantum system with its environment. In the Schrödinger's cat thought experiment the interaction of a particle with the detector will cause it to decohere – meaning the cat is definitely alive or dead long before the box is opened.

the waves spread out in a characteristic pattern as they emerge. Sure enough, the team saw the narrow electron beam fan out in a diffraction pattern on the far side of the grid. Electrons – particles of matter – were being diffracted, an inherently wave-like behaviour – exactly as predicted by de Broglie.

The nature of the connection between waves and particles became clearer when an Austrian physicist called Erwin Schrödinger heard of de Broglie's ideas. Schrödinger took standard relations between energy and momentum from Newtonian mechanics and plugged in Planck's and de Broglie's expressions for these quantities in terms of waves. As he expected, the result, which he first wrote down in 1926, was a wave equation.

Schrödinger's equation predicted the observed structure of the hydrogen atom – a feat that no other theory had accomplished. But he was still baffled as to what exactly his equation meant. It used a nebulous quantity that Schrödinger called the 'wave function' to infer the values of physical quantities, but what exactly did the wave function itself represent?

THE BORN REVELATION

It was German physicist Max Born who ultimately figured out the solution to this mystery. Schrödinger's equation for a particle predicts the way the particle's wave function varies with position, and Born realized that taking the square of the wave function (multiplying it by itself) gave the probability of finding the particle at any given point in space. Schrödinger's equation meant that it was no longer possible to predict a particle's location with certainty, as it was in Newtonian mechanics – quantum theory simply gave the likelihood that the particle might be found at a certain location when a measurement is made.

This is how an electron can be both a particle and a wave at the same time – the probability of the electron's location behaves like a wave, and only when a measurement is made do we see the electron as a particle with a definite position in space. Ironically, Einstein – who had sown the seeds of quantum theory – came to detest this apparent randomness, famously declaring: 'God does not play dice with the universe.' Nevertheless, the results have stood up to rigorous experimental scrutiny, and applications of quantum mechanics have given rise to technological advances from lasers and LEDs to medical imaging techniques, advanced encryption and nuclear power.

The condensed idea
When entering the subatomic world, leave Newton at the door

09 Quantum fields

Applying Einstein's ideas of relativity to quantum mechanics leads to a quantum picture for the behaviour of 'fields', entities that explain how the forces of nature exert their influence over matter. The quantum theory of the electromagnetic field remains one of the most accurate theories ever built in physics.

For all the successes of quantum mechanics (see page 32), the theory took no account of the other great scientific revolution of the 20th century: relativity. Einstein's special theory of relativity says that the laws of motion for objects travelling at close to light speed are very different to those of our everyday experience. And so, as it stood, quantum mechanics was unable to describe fast-moving particles.

British physicist Paul Dirac changed that when, in 1928, he reformulated Schrödinger's wave equation for the behaviour of a quantum particle to make it consistent with special relativity. The result was the Dirac equation, a quantum wave equation for the motion of high-speed electrons (negatively charged particles that orbit in the outer regions of atoms).

ANTIMATTER

It soon became clear that Dirac's equation had some surprises in store. It naturally explained the concept of 'quantum spin' – a kind of quantum analogue to everyday rotation, which had been introduced to explain peculiarities in the behaviour of electrons by the Swiss physicist Wolfgang Pauli. Dirac's equation admitted not only a mathematical solution

TIMELINE

1928	1932	1948
Dirac develops the first theory unifying quantum mechanics and special relativity	Carl Anderson discovers antimatter, predicted by Dirac's theory	Feynman, Schwinger and Tomonaga formulate quantum electrodynamics, also known as QED

describing the electron, but also a second solution corresponding to a particle with the same mass as the electron but opposite (positive) electric charge. The particle, called the positron, was the first example of antimatter: it was duly detected by American physicist Carl Anderson in 1932.

FORCE FIELDS

Dirac's equation described not only the motion of electrons and positrons, but also the interactions between them owing to their electric charges. Previously, this area of physics had been governed by James Clerk Maxwell's classical theory of electromagnetism, which predicted how electric charges generate electric fields, and how other charges move under the action of these fields. Now, Dirac's equation replaced classical electromagnetism with a quantum description. It was the first 'quantum field theory'.

Planck and Einstein (see page 32) had already established the photon as the particle associated with electromagnetic waves. But Dirac's work took this idea to a new level. Every time an electrically charged particle interacted with an electromagnetic field, it was doing so through the exchange of photons. In this regard, photons are the 'force carriers' of the electromagnetic field. It became clear that photons can spontaneously bubble up from the background field. Pairs of these so-called 'virtual photons' can pop into existence for short

Heisenberg's uncertainty principle

The uncertainty principle is a fundamental tenet of quantum theory, concerning the precision with which the properties of quantum particles can be known. It was put forward in 1927 by the German physicist Werner Heisenberg.

In its basic form, the principle says that the uncertainty in a particle's position multiplied by the uncertainty in its momentum is always greater than or equal to Planck's constant divided by 4π, where Planck's constant is 6.63×10^{-34} (see page 32). So if you reduce the uncertainty in one quantity then the other must increase.

An alternative form of Heisenberg's uncertainty principle says that energy and time obey a similar relation to position and momentum. This explained the creation and annihilation of particles in quantum field theory.

1954

Yang and Mills develop gauge theories, which underpin modern quantum fields

1972

Fritzsch, Gell-Mann and Leutwyler develop quantum chromodynamics, or QCD

1979

Glashow, Salam and Weinberg win the 1979 Nobel Prize for the electroweak model periods of time, in accordance with the uncertainty principle (see box on page 37). The principle allows the field to borrow the energy needed to create a particle pair, provided the pair annihilate and return the energy a short time later: the more energy borrowed, the shorter the life of the particles.

> **ANYTHING YOU CAN DO IN CLASSICAL PHYSICS, WE CAN DO BETTER IN QUANTUM PHYSICS.**
> Daniel Kleppner

TROUBLESOME INFINITIES

Nevertheless, Dirac's theory was flawed. For some physical quantities it gave absurd infinite answers. And it failed to predict a small difference between two energy states of the hydrogen atom – known as the Lamb shift after its discoverer, American physicist Willis Lamb. It wasn't until the late 1940s that solutions to these issues were found. To fix the infinities, physicists Julian Schwinger and Sin-Itiro Tomonaga independently proposed a scheme known as 'renormalization'. When performing calculations in quantum theory it's often impossible to solve the mathematical equations exactly. Instead, physicists use a 'perturbative expansion', a series of terms of increasing complexity that together offer an approximation to the exact solution. Renormalization amounted to simply discarding the terms in the expansion that were causing the infinities. It sounds very ad hoc, but modern renormalization theory gives a physical interpretation to the process: parameters in the theory can vary with the energy at which observations are made.

Meanwhile, at the California Institute of Technology, physicist Richard Feynman had realized that a particle moving between two points could take any of a number of different routes. To calculate the total probability of the particle making the journey, Feynman added up the probability of each possible route, an approach that he called the 'path integral' formulation of quantum theory.

FEYNMAN DIAGRAMS

Feynman supplemented his technique with a way of visualizing it pictorially. This amounted to drawing a diagram for every possible way that a quantum interaction could proceed. Each of these Feynman diagrams implied a particular mathematical contribution that could then be added to the path integral. When Feynman applied these developments to electromagnetism, the Lamb shift was now present and its magnitude matched exactly with experimental observations. The theory was called quantum electrodynamics or QED, and it was so successful – with predictions accurate to an astonishing 11 decimal places – that it won Feynman, Schwinger and Tomonaga the 1965 Nobel Prize in Physics.

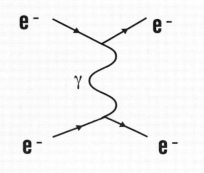

This Feynman diagram shows two electrons scattering off one another through the exchange of a photon. Here, time increases from left to right.

Today, we have more quantum field theories describing the other forces of nature on the smallest scales imaginable. Quantum chromodynamics (QCD) was put forward in 1972 as an explanation of the strong nuclear force, which binds quarks together to make neutrons and protons, while the 'electroweak' model, devised in 1967, combines QED with a theory of the weak nuclear force (which is responsible for radioactive decay – see page 76). Both have been verified by rigorous experimental tests. The hunt is now on for a scheme to combine QCD and the electroweak model into one 'grand unified theory' (see page 48).

QUANTUM GRAVITY

The one field that has thus far refused to be quantized is gravity – with attempts suffering from infinities that cannot be removed by renormalization. Our best classical (ie, non-quantum) theory of gravity is general relativity (see page 28), which ascribes the gravitational force to curvature of space and time rather than particles and fields existing within it. It may be that this fundamental difference means that we will never see a full quantum theory of gravity. Only further research will tell.

The condensed idea
Action at a distance also obeys quantum laws

10 Particle physics

In 2012, physicists at the Large Hadron Collider discovered the long-sought Higgs particle. Its detection completed the 'standard model' – our best theory of subatomic particles and the forces between them. Now scientists are pushing further to investigate the particle world in which the universe was born.

The standard model of particle physics was drawn up in the early 1970s. It followed from a number of breakthroughs in quantum field theory (see page 36) – in particular, the development of quantum electrodynamics (QED) in the 1940s, the electroweak theory of the 1960s, and quantum chromodynamics (QCD) in 1972. Scientists were finally in a position to pull together everything they knew about the particle world into one big picture.

IN A SPIN

The model splits the known particles into two broad families, called fermions and bosons, according to their 'quantum spin'. This is a property analogous to, but rather different from, spin in the everyday sense. Every different particle species has a fixed spin, describing its symmetry when rotated. For example, a particle with spin 1 will look the same after one complete rotation. A particle with spin 1/2 must be rotated fully twice to get back to its starting state. Bosons have whole-number spin (0, 1, 2, etc.), while fermions have half-whole-number spin (1/2, 3/2, 5/2, etc.).

The fermion family breaks down further into hadrons (particles that feel the strong nuclear force) and leptons (those that do not). The lepton class

TIMELINE

1964
The quark theory is proposed by Murray Gell-Mann and George Zwieg

1974
John Iliopoulos presents the 'standard model' as such for the first time

1983
The W and Z particles of the electroweak theory are detected

includes electrons – the particles we met earlier, which inhabit the outer regions of atoms – plus two sister particles, called the muon and the tauon. All three particles have electric charge of -1 and spin of 1/2. However, their masses are very different. The muon weighs 200 times more than the electron, while the tauon is 17 times heavier again. Each of these three lepton types comes with its own 'neutrino' – a ghostly particle that has no electric charge, almost no mass and spin 1/2. Each of the six lepton particles also has its own antiparticle (a particle with the same mass and spin, but with other key properties reversed).

QUARK, STRANGENESS AND CHARM

The other class of fermions are the hadrons. At the most fundamental level, these are quarks. They come in six 'flavours', labelled: up (u), down (d), strange (s), top (t), bottom (b) and charm (c) and all have spin 1/2. They also have fractional electric charge of either +2/3 or -1/3. The quark model states that the familiar protons and neutrons that make up the nuclei of atoms are made from clusters of three quarks. For example, the proton is made of two 'u' quarks (charge +2/3 each) and a 'd' quark (charge -1/3), to give an overall charge of +1.

Particle accelerators

Theories in particle physics are tested in large machines called particle accelerators. These use a series of magnets to accelerate electrically charged subatomic particles to near light speed and smash them into one another. The debris from the collisions can then be examined for the signatures of rare particle species.

We more or less understand how the present-day particle world operates. Most of the unknowns lie in the past when the universe was much hotter and denser, and probing these regimes requires accelerators of ever-increasing size and power.

The world's most powerful particle accelerator today is the Large Hadron Collider at CERN in Geneva. Its latest upgrade, completed in 2015, gave the accelerator double the energy of the incarnation that discovered the Higgs particle in 2012.

Quarks also carry 'colour charge'. This has nothing to do with actual colour – the name is just a label. Whereas electric charge acts as a source for the electromagnetic force, colour charge is the source for the strong nuclear force

1995
The elusive top quark is discovered by scientists at Fermilab

2000
The tau neutrino is found by Fermilab's DONUT experiment

2012
The Higgs boson is finally detected at the Large Hadron Collider

that binds together quarks to form protons and neutrons, and in turn the nuclei of atoms. The theory of quantum chromodynamics (QCD) describes three types of colour charge (labelled 'red', 'green' and 'blue'), which, like electric charge, can be either positive or negative. As well as protons and neutrons, which are sometimes collectively known as 'baryons', quark-antiquark pairs, known as 'mesons', can also form. Lone quarks are never seen, however.

> **I HAD A BET WITH GORDON KANE OF MICHIGAN UNIVERSITY THAT THE HIGGS PARTICLE WOULDN'T BE FOUND. THE NOBEL PRIZE COST ME $100.**
> Stephen Hawking

The quark model was proposed by Murray Gell-Mann and George Zweig, working independently, in 1964. All six quark flavours have now been observed, with the last one, the top quark, emerging in 1995 – at the Tevatron particle accelerator, at Fermilab in Chicago.

FORCE CARRIERS

Quantum field theory states that the forces between particles of matter are mediated by 'exchange particles' of the field. These all have spin equal to 1, and so occupy the boson section of the standard model. For example, in electromagnetism, the exchange particle is the photon.

In quantum chromodynamics, the exchange particles are called 'gluons', and they come in eight different types. The mathematics of QCD requires each gluon type to carry a different mixture of positive and negative colours. Meanwhile, the weak nuclear force is mediated by the exchange of particles called W and Z. The Z is electrically neutral, but the W comes in two types with electric charges +1 and -1. The W and Z particles were observed experimentally in 1983, with masses almost identical to theoretical predictions.

But there's one other boson particle in the standard model. Scientists at the Large Hadron Collider particle accelerator at CERN, on the Swiss-French border, grabbed headlines in 2012 when they announced the discovery of the elusive Higgs boson.

THE HIGGS BOSON AND BEYOND

In order to explain the masses of other particles in the standard model British physicist Peter Higgs and others proposed the existence of an uncharged, colourless, spin-zero particle in 1964. The Higgs boson is the particle of a quantum field, called the Higgs field, that pervades all of space. Particles interacting with this field acquire mass – causing them to be slowed down, like a spoon moving through treacle. Prior to its introduction, the standard model had no way to predict the masses of its constituent particles.

The Higgs was initially proposed as part of the electroweak model, a quantum field theory that unifies electromagnetism and the weak nuclear force at high energy – for example, shortly after the Big Bang, in which the universe was created (see page 172). At the lower energies of the present day, electromagnetism and the weak force exist as distinct entities.

The next challenge facing particle physics is to repeat this feat of unification for the electroweak and strong nuclear forces, joining them together in a so-called 'grand unified theory', or GUT. There are already many candidate theories: the problem is distinguishing between them. Testing such a model would require a particle accelerator a trillion times more powerful than the 27-kilometre (17-mile) diameter Large Hadron Collider.

Symmetry

Much of modern particle physics is built on the concept of symmetry. In classical physics, symmetries are connected to 'conserved quantities'. For example, in Newtonian mechanics, the law of conservation of energy – that energy can be neither created nor destroyed – follows from demanding time symmetry. This simply says that the laws of physics tomorrow should be the same as they are today.

In 1915, German mathematician Emmy Noether proved that conserved quantities in any physical theory correspond to its symmetries. For example, electromagnetism has one symmetry, meaning there should be one conserved quantity. And indeed there is – electric charge. In contrast, quantum chromodynamics, the theory of the strong nuclear force, has three symmetries – one for each of the three types of colour charge.

Physicists classify the symmetries of different particle theories using a branch of mathematics called group theory.

The condensed idea
The hidden order in the fundamental particles of matter

11 Nuclear energy

Quantum theory has given us electronic devices from Blu-ray players to computers. But it has also provided the electricity to run these devices through nuclear power, extracting energy from atoms. Despite some high-profile accidents, experts believe that nuclear power remains the cleanest practical energy source.

In 1911, New Zealand-born physicist Ernest Rutherford made a startling discovery. Working with colleagues at the University of Manchester, he found that most of the mass of an atom is concentrated into a central 'nucleus' very much smaller than the atom itself. In fact, if an atom was scaled up to the size of a football pitch then the nucleus would be roughly the size of a pea.

> **NOW I AM BECOME DEATH, THE DESTROYER OF WORLDS.**
> Robert Oppenheimer, on witnessing the first atom bomb test

Rutherford followed up the breakthrough in 1919 by demonstrating that the nucleus isn't a solid lump but is itself built up from smaller particles called protons, each of which carries a single unit of positive charge. But this brought a new mystery – why do large atomic nuclei, which are made of many protons, not fly apart thanks to the electrostatic repulsion between all those positive charges?

This problem was solved in 1932 by the British physicist James Chadwick, when he discovered a new subatomic particle with the same mass as the proton but zero electric charge. Chadwick had found the neutron, a particle that sits in between protons in the nucleus, moderating the force between them and thus helping hold the nucleus together.

TIMELINE

1911	1919	1932
Ernest Rutherford, at the University of Manchester, discovers the atomic nucleus	Rutherford and colleagues find that the nucleus contains proton particles	Rutherford's former student James Chadwick discovers the neutron

SPLITTING ATOMS

In 1938, German physicists Otto Hahn and Fritz Strassmann were conducting experiments to bombard the heavy element uranium with Chadwick's neutrons. They hoped that some of the neutrons might be captured by uranium nuclei and then change into protons via radioactive beta decay (see page 76). Because a chemical element is defined by the number of protons in its nucleus (see page 72), Hahn and Strassmann were essentially trying to transmute uranium into another element – in this case, the heavier element neptunium.

However, to their bafflement, they actually ended up creating small amounts of the very much lighter elements barium and krypton. It was Austrian physicists Lise Meitner and Otto Frisch who cleared up the mystery. While a few neutrons are needed to hold the nucleus together, too many make it unstable. Meitner and Frisch realized that the lighter elements must be formed by uranium nuclei splitting apart. They called the process 'nuclear fission', a nod to the process of 'binary fission' describing the division of biological cells.

Nuclear binding energy

Why is it that fission (splitting heavy atomic nuclei apart) and fusion (joining lighter ones together) both release energy? You might think it would be either one or the other.

The answer to this question is all to do with what's called 'nuclear binding energy'. This is the energy that you have to add to an atomic nucleus to break it apart. Conversely, when the nucleus is formed from disparate protons and neutrons, this amount of energy must be released.

The binding energy of different elements is predicted by the strong nuclear force, which is what ultimately holds the nucleus together. Plotted on a graph, the binding energy steadily rises as atoms get more massive, reaching a maximum value at iron, before steadily tailing off again for very heavy elements.

And this is why splitting elements heavier than iron, and fusing together those lighter than iron, both raise the overall binding energy of the product nuclei, causing energy to be given off.

The combined mass of barium, krypton and the remaining uranium was significantly less than the initial mass of uranium. It was Meitner who

1938
German and Austrian physicists demonstrate nuclear fission for the first time

1945
The first atomic bombs used in warfare kill nearly 200,000 Japanese civilians

1986
World's worst nuclear disaster, at the Chernobyl Nuclear Power Plant, Ukraine

pointed out that, because of Einstein's famous equation $E = mc^2$, this deficit in mass must correspond to the release of a large quantity of energy. And it wasn't just energy being given out. Each time a nucleus splits, it's accompanied by a burst of neutrons. And these neutrons can be absorbed by other uranium atoms, causing them to split, and so on, and so on. The concept of a 'nuclear chain reaction' was born.

THE FIRST REACTORS

In 1942, this theory was put into practice for the first time. A team led by Italian-American physicist Enrico Fermi built the world's first artificial nuclear reactor in a disused squash court at the University of Chicago. The reactor consisted of a pile of raw uranium surrounded and interspersed with graphite blocks to soak up excess neutrons. In addition, a set of neutron-absorbing control rods enabled scientists to dictate the rate of the reaction, withdrawing the rods to speed it up. On 2 December, 1942, the reactor produced enough neutrons to sustain a chain reaction.

Chicago Pile-1, as it was known, formed an early stage in the Manhattan Project – America's effort to build the first atomic bomb. Although the effort was motivated by fears of Nazi Germany developing such weapons, Hitler's forces had already surrendered by the time the first bomb was tested. Instead, the first live target of nuclear weapons was Japan, with the atomic bombings of Hiroshima and Nagasaki on 6 and 9 August, 1945. Following the war, however, nuclear fission was turned to more peaceful uses.

Whereas nuclear fission involves the splitting of heavy atomic nuclei, fusion involves the bonding together of lighter ones.

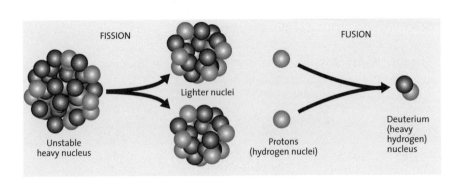

FISSION

FUSION

Lighter nuclei

Unstable heavy nucleus

Protons (hydrogen nuclei)

Deuterium (heavy hydrogen) nucleus

The world's first commercial nuclear power plant opened at Calder Hall, in England, in 1956. Many more have been commissioned since. There are currently estimated to be over 430 civil nuclear reactors in operation, providing around 11 per cent of the world's energy. Nuclear power hasn't been without its safety concerns, but some scientists, including prominent British environmentalist James Lovelock, argue that it is the most environmentally friendly power source we have. They say that occasional high-profile accidents, such as Chernobyl in 1986 and Fukushima in 2011, are still nowhere near as polluting as the day-to-day operation of fossil-fuel power plants. Meanwhile, renewable sources, such as solar and wind power, may be unable to meet demand on their own.

FUSION RECIPE

One future hope is that nuclear fission power may one day be replaced with the much cleaner nuclear fusion. Rather than splitting heavy elements, fusion works by joining together lighter ones, such as hydrogen, giving off energy in the process. It's the same physics that powers the Sun, and requires extremely high temperatures to ignite (millions of degrees Celsius). Unlike fission, however, the waste products from nuclear fusion are not radioactive.

Although successfully demonstrated in advanced nuclear weapons (so-called thermonuclear or hydrogen bombs), controlling a fusion reaction to the degree needed for power generation is a feat that has thus far eluded scientists. The best efforts to date use magnetic fields to confine the positively charged hydrogen nuclei while they are heated. Most experts agree that the present rate of progress is unlikely to yield commercial nuclear fusion reactors much before the year 2050. However, as attention focuses ever more urgently on how to keep the lights on without choking the planet, fusion power may well be a technological development that we postpone at our peril.

The condensed idea
Awesome power from little atoms grows

12 String theory

Could it really be that the most promising candidate for a unified theory of all the fundamental forces of nature is based on the premise that all the matter in the universe is made from string? Many physicists believe so, though of course these are no ordinary strings, but vibrating strings of energy.

In the 19th century, Scottish physicist James Clerk Maxwell was able to combine the theories of electricity and magnetism into one unified entity, known as electromagnetism (see page 16). The new theory was much bigger than the sum of its parts – leading to new insights into the behaviour of light and paving the way for Einstein's ideas on relativity. This idea of unification in physics is admired for its elegance and ultimate simplicity – explaining the laws of the universe from as few fundamental assumptions as possible.

A THEORY OF EVERYTHING

In the 1970s, physicists were able to unify electromagnetism with the weak nuclear force (one of two forces presiding inside the nuclei of atoms, responsible for radioactivity – see page 76). There are also a number of candidate theories building the strong nuclear force into this scheme.

But there's a fourth and final force of nature – gravity. Slotting this into a unified scheme with the other three forces has proven to be a mathematical nightmare. The weak and strong nuclear forces exist solely within the nuclei of atoms and so any unification scheme incorporating them must embrace quantum principles. But gravity's not having any of it. Not only do some

TIMELINE

1921	1968	1981
Theodor Kaluza publishes a unification model requiring 5D spacetime	Gabriele Veneziano develops the first incarnation of string theory	Superstring theory is formulated by Michael Green and John Schwarz

physical quantities in the candidate models tend to infinity, but these 'divergences' cannot be fixed by renormalization, a technique that was used to remove similar issues from the quantum theory of electromagnetism (see page 38).

Traditional quantum theory treats fundamental particles of matter as points with zero size, and some physicists speculated that this may be the cause of the divergences. In reality, a particle, no matter how small, will always have some finite extent. Packing the particle's mass into zero volume instantly makes its density infinite – and this, reasoned the physicists, could be the source of the divergences in quantum gravity.

> GOOD WRONG IDEAS ARE EXTREMELY SCARCE, AND GOOD WRONG IDEAS THAT EVEN REMOTELY RIVAL THE MAJESTY OF STRING THEORY HAVE NEVER BEEN SEEN.
> Edward Witten

String theory gets around this by supposing that particles aren't points but tiny, one-dimensional 'strings' of energy. They really are small – around 10^{-33} cm long (that's a decimal point followed by 32 zeros and then a 1). That's so small that if an atom was blown up to the size of the observable universe then a fundamental string would be about the same size as a tree. The strings vibrate and the frequency of the vibration dictates the particular type of particle that each particular string represents – rather like different notes being played on a guitar string.

HIGHER DIMENSIONS

Mind-bogglingly, in order to keep the mathematics of string theory self-consistent, space and time are required to have ten dimensions – six more than the four (three of space and one of time) that we can actually see. The theory has it that these extra dimensions are 'compactified' – tightly curled up so as to make them invisible. It's a bit like the way the surface of a hose pipe, which is actually a two-dimensional cylinder, appears as a one-dimensional line when viewed from a distance. In string theory, however,

1983
Witten and Alvarez-Gaume build the first superstring quantum gravity theory

1991
Veneziano shows how string theory could have ruled the early universe

1995
Witten pulls together different versions of string theory to create M-theory

nothing is quite that simple, and to satisfy the constraints of the theory, the compactified dimensions have to be wrapped up into a complex knot of spacetime called a Calabi-Yau manifold.

The first incarnation of string theory was devised in 1968, by Italian theoretical physicist Gabriele Veneziano. He was actually trying to build a model of the strong nuclear force, and found that a string-like approach was able to explain many of the observed characteristics of strong interactions. Although Veneziano's theory of the strong force was ultimately overtaken by quantum chromodynamics (see page 39), interest continued throughout the 1970s. In the 1980s, American physicist Edward Witten showed that string theory did indeed lead to a workable quantum theory of gravity – neatly eliminating the divergences that had plagued early attempts. Others have since used string-inspired models to investigate mysteries in the physics of black holes, and to ask what may have happened before the Big Bang in which our universe was created.

Nevertheless, string theory has drawn criticism from some quarters. Although mathematically self-consistent and non-divergent, the theory itself is yet to be properly 'defined'. Ordinarily in particle physics, scientists do calculations using what's called 'perturbation theory'. This uses the technique of Feynman diagrams (see page 38) to make an approximation to the full theory, which is usually too complicated to solve exactly. In building string theory, they simply replaced the particles in perturbation theory with strings – giving a new approximation. The trouble is that no one quite knows what it's an approximation of. Others quibble that testing string theory could present problems. That's because the energy scale

 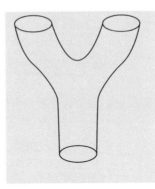

The Feynman diagram for two interacting point particles (far left), and the same diagram with the particles replaced by loops of string (near left). In these diagrams, time runs from top to bottom.

at which gravity gets incorporated into unified models is way beyond the reach of even our most powerful particle accelerators. More recently, however, it's been suggested that the precise form of the compactification of spacetime may leave an observable signature that we can look for, and which could form the basis of an experimental test.

M-THEORY

In 1995, Witten took string theory to a whole new level when he introduced the concept of M-theory. String theory is in fact a number of different theories, depending how the parameters in the model are chosen, and M-theory pulls all of these together into a single consolidated picture. In M-theory, one-dimensional strings in ten-dimensional spacetime are replaced by two-dimensional 'membranes' in eleven-dimensional spacetime. Any particular string theory then corresponds to a particular 'slice' through M-theory.

Superstrings

In 1983, physicists Michael Green, from England, and John Schwarz, from the United States, were able to marry string theory with supersymmetry – a framework in particle physics believed to be essential for the unification of the forces. It says that every fermion (particle with half-integer spin) has a partner that's a boson (particle with integer spin). There is no direct evidence for supersymmetry – it's motivated on largely aesthetic grounds, and from the pivotal role played by conventional symmetries in the standard model. The resulting 'superstring theory' became the first working quantum theory of gravity.

String theory and M-theory have spent many decades in development without quite delivering the grand answers that everyone had hoped for. But many remain optimistic that these theories may yet lead to a testable unification model bringing together all four of the forces of nature. This is new physics at the very boundaries of our understanding, and as the history of science has taught us time and again, pushing back those boundaries takes ingenuity, inspiration – and time.

The condensed idea
Matter on the smallest scales is made of string

13 Information theory

The mathematical theory governing the behaviour of information may sound like a very abstract concept indeed, but today it is instrumental in communications technology, computers and data analysis. It has also played a role in understanding black holes – and it's the reason why a scratched CD will still play.

Up until the 1940s, no one had a clue quite what information was, let alone how to quantify it mathematically. That all changed with the work of an American electronics engineer and mathematician called Claude Shannon. During the Second World War, Shannon had worked at Bell Labs, New Jersey. He spent time developing ideas in cryptography (briefly collaborating with British codebreaker Alan Turing), and worked on the fire-control systems for artillery guns. In particular, he looked at the case when there was noise hindering the communication channel between the controller and the gun – and developed methods for minimizing the effects of the noise, and thus maximizing the efficiency with which signals could be transmitted.

The subject piqued Shannon's interest and he continued his research in this area after the war, leading to the publication in 1948 of a seminal book, entitled *A Mathematical Theory of Communication*. With it, the science of information theory was born.

BIT WISE
The first thing Shannon did was to define the slippery concept of information in terms of rigorous mathematics. He did this by introducing the 'bit' as the

fundamental unit. A bit was a way to encode the simple on/off state of a switch and, as such, could take the values 0 (off) or 1 (on). Two bits working together could encode twice as many numbers – 0 (0-0), 1 (0-1), 2 (1-0) or 3 (1-1). The set of bits making a number was called a 'byte'. Adding more bits to the byte extended the range of numbers that it could store. In general, an n-bit byte can store all numbers from 0 to $(2^n - 1)$. So, today's latest computers, which are 64-bit machines, are essentially able to count to over 18 billion billion. Interestingly, your two hands form a 10-bit byte, enabling you to count from 0 to 1023 on your fingers – much better than just 10!

THERE WERE MANY AT BELL LABS AND MIT WHO COMPARED SHANNON'S INSIGHT TO EINSTEIN'S. OTHERS FOUND THAT COMPARISON UNFAIR – UNFAIR TO SHANNON.
William Poundstone

SOURCE CODE

With the concept of the bit in place, Shannon began to consider how to transport bits most efficiently between a transmitter and a receiver. One idea he developed was known as 'source coding', which amounted to paring down the number of bits in a signal to its absolute bare minimum. For example, let's say you flipped a coin 1,000 times and wanted to send a message giving the state of every flip. Each flip has a basic head/tail outcome. We could encode this using one bit for each flip (for example, setting 1 = heads and 0 = tails), so ordinarily it would take 1,000 bits to encode the message. But let's say the coin is biased somehow, so that its odds of coming up heads are only 1 in 1,000. Now, on average, there will only be one head in our 1,000 flips. So all we need to transmit is the position of this head in the sequence, and we can do that with just 10 bits (using the formula above, $2^{10} - 1 = 1,023$) – much more economical than transmitting the individual state of all 1,000 bits.

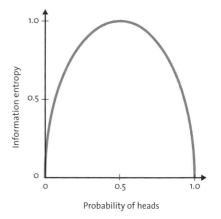

The information entropy of a flipped coin reaches a maximum when the probability of heads is 0.5, or 50 per cent. This is the state with maximum uncertainty.

Claude Elwood Shannon (1916–2001)

Claude Shannon was born on 30 April 1916 in Michigan to Mabel, a languages teacher, and Claude Sr, a businessman. As a child he had a keen interest in electrical and mechanical devices, one time building an electric telegraph system between his and a friend's house.

Shannon entered the University of Michigan in 1932, graduating in 1936 with degrees in mathematics and electrical engineering. Following this, he enrolled as a graduate student at MIT, before moving to Princeton's Institute of Advanced Study in 1940.

As an avid hobby inventor, he built devices including a rocket-powered Frisbee, and a machine to solve the Rubik's Cube. He also co-designed the world's first wearable computer – to help him cheat at roulette. He met his wife Betty while at Bell Labs. They married in 1949 and had three children together. Claude Shannon died in 2001 at the age of 84.

Shannon devised a measure for how much a message could be compressed. The higher the uncertainty in the content of the message, the less compressible it would be – and the more bits it would take to transmit. He was stumped over what to call the new property until his friend the Hungarian-born mathematician John von Neumann suggested the word 'entropy', borrowed from the quantity in thermodynamics that characterizes the degree of 'disorder' in physical systems (see page 20). It's very much like Shannon's uncertainty, and can be calculated using a similar formula.

DATA DUMP

Every time you 'zip' large files on your PC, such as photos or videos, you are making use of Shannon's concept of information entropy to strip out all of the redundant data and minimize the space taken up by the files – for instance, before you upload them to a backup server online. This is called 'lossless' compression because no information is lost in the process. Other forms of data compression are 'lossy' because information is actually being discarded. Modern audio compression standards work by throwing away data corresponding to frequencies that are beyond the limits of human hearing. This enables a typical MP3 music file to be about 9 per cent of the size of the same song at full quality.

Another major contribution by Shannon was the concept of 'channel coding' to try and gauge the amount of noise in a communication channel and then correct for it. The technique involves the message sender splicing a sequence of bits, pre-agreed with the receiver, into the message before sending. The receiver can then observe how this known sequence has been scrambled, in order to figure out the form of the noise and subtract it out from the rest of the message. Channel coding has given us modern error-correction methods – which allow you to have an audible conversation over a crackly phone connection, listen to a scratched CD or scan the barcode on a scrunched up bag of sweets.

INFORMATION AGE

Today, information theory is essential in communications technology, computer science and data analysis. It contributes to security – helping us to make and break codes, and to detect criminal and even terrorist activity.

The Kelly criterion

In 1956, American mathematician John Kelly published a scientific paper in the *Bell System Technical Journal* with the rather bland title, 'A New Interpretation of Information Rate'. In it, Kelly used information theory to derive a formula for the fraction of a gambler's bankroll that they should bet on each of a series of wagers in order to win the most money.

Kelly found that if the actual probability of the bet winning is p (between 0 and 1) and the odds offered by the bookmaker are b/1 then the optimal fraction to bet is [p × (b+1) -1] /b. If it's negative you don't bet.

The so-called Kelly criterion has been successfully applied by gamblers, as well as investors betting in the biggest casino of them all – the stock exchange.

Information theoretic treatment of randomness and uncertainty provides significant insights in gambling and investing. Meanwhile, it is also playing a role in other, more esoteric branches of science. It's helping to explain how genes organize themselves inside biological cells. The theoretical study of how information can potentially leak out of a black hole has stimulated new ideas in quantum gravity, in particular string theory (see page 48). And this in turn has led to some radical suggestions in cosmology, such as the extraordinary notion that the universe could actually be a hologram.

The condensed idea
The mathematical laws of communication

14 Chaos theory

If you balance a pencil on its end and let go, which way will it fall? The tiniest shift in its state of balance could send it one way, or in the other direction entirely. The physics describing the pencil's behaviour is well understood, and yet we still have no idea what's going to happen. This is an example of chaos.

Some things in nature are so sensitive to their starting state that predicting their future behaviour is next to impossible – even when the system itself is well understood. These are chaotic systems – phenomena in nature that are 'deterministic' but that nevertheless move apparently at random. The phenomenon crops up exclusively in systems that are governed by non-linear maths. Linear equations are simple relationships like $y = 2x$. If you increase x, then y increases in proportion. For instance, change x from 1000 to 1000.1 – a tiny change of 0.1 – and y changes by 0.2, which is no big deal. Non-linear equations, on the other hand, are more complex – sometimes deceptively so. An example is $y = x^2$. Now increase x, and y increases by a disproportionate amount – change x from 1000 to 1000.1, and the relatively small shift in x now produces a huge shift in y of more than 200.

Non-linear equations can lead to large changes in behaviour from very small changes in the starting conditions, and that's why we've no idea which way the pencil will fall. It's not so much the mathematics that's responsible, but our inability to measure the tiny variations in its starting state accurately enough – and these tiny variations produce large effects.

TIMELINE

1880s	1961	1972
Poincaré finds the signs of chaos theory in Newtonian gravity	Edward Lorenz discovers chaotic behaviour in weather systems	Lorenz coins the 'butterfly effect' to describe sensitivity of chaotic systems

It was the French mathematician Henri Poincaré, during the 1880s, who first tried to quantify the effects of chaos in the real world. Studying Newton's law of gravity (see page 12), he found that while solutions are well behaved and predictable in the case of two objects (such as planets) moving in each other's gravitational fields, the addition of a third body made the future behaviour of the system extremely hard to predict.

> **CHAOS: WHEN THE PRESENT DETERMINES THE FUTURE, BUT THE APPROXIMATE PRESENT DOES NOT APPROXIMATELY DETERMINE THE FUTURE.**
> Edward Lorenz

WEATHER WATCHING

Despite Poincaré's work, it took until the second half of the 20th century for his findings to gain traction. In 1961, American mathematician Edward Lorenz was carrying out a study of convection in weather systems – circulating currents in the atmosphere caused by rising hot air – using an early digital computer. At one stage in his research, Lorenz needed to recreate some output that he had produced earlier. Rather than running the computer program all the way from the beginning, he re-entered data from a printout taken at the midpoint of the calculation, and then set the computer to work. To his astonishment, he found that the results of the calculation – the weather patterns that it predicted – were completely at odds with what he had seen on the first run.

Lorenz eventually figured out what the problem was. During the calculation, the computer had been storing data to six decimal places, but the printout truncated these numbers to just three decimal places. So, for example, the number 0.437261 was being logged on the print-out as 0.437. This tiny difference was responsible for the huge difference in the final results: Lorenz was seeing chaos in action. Chaos is now a well-known feature of weather systems, and is the reason why forecasters can never predict more than a few days in advance.

1975
The term 'chaos theory' is coined by James Yorke and Tien-Yien Li

1982
Mandelbrot publishes his seminal book *The Fractal Geometry of Nature*

1984
NASA's Voyager 2 reveals the chaotic rotation of Saturn's moon Hyperion

Edward Norton Lorenz (1917–2008)

Edward Lorenz was born on 23 May 1917 in Connecticut, USA. He studied mathematics, first at Dartmouth College, New Hampshire, earning a bachelor's degree in 1938, and then at Harvard, obtaining a master's in 1940.

During the Second World War, he served as a meteorologist with the US Army Air Corps. The subject interested him and after the war he enrolled to study it at the Massachusetts Institute of Technology (MIT), where he obtained a doctorate in 1948. He remained at MIT for the rest of his career, becoming a full professor in 1962.

The 1950s and 1960s saw the advent of digital computers in science, which were a major new resource with which to try and unpick the complexity of weather systems. Lorenz was among the first to do this – and it was this research that led him to discover chaos theory. Lorenz was married and had three children. He died in Cambridge, Massachusetts, on 16 April 2008.

Lorenz published his research in 1963. He would later coin a term that has become synonymous with chaos: the 'butterfly effect'. It refers to the notion that a butterfly beating its wings could alter the state of the weather sufficiently to potentially trigger a tornado on the other side of the world. The phrase originally featured in the title of a talk by Lorenz at the annual meeting of the American Association for the Advancement of Science in 1972.

Scientists can now look for chaos in a system using a scheme originally developed by the Russian mathematician Aleksandr Lyapunov in 1892. Although unaware of chaos per se, Lyapunov worked on the stability of the solutions to non-linear equations. This involved making small perturbations to a known solution of the equations. If the perturbations die away over time then the solution is deemed 'stable'. If they grow larger over time then the solution is 'unstable'. This test for instability – quantified by a set of numbers called the 'Lyapunov exponents' – is now regarded as the acid test for detecting chaotic behaviour.

STRANGE ATTRACTORS

There's also a geometrical twist to chaos theory. The motion of an object can be characterized by what's called a 'phase portrait' – essentially a graph of the object's speed against its position. For example, motion at constant speed is just a straight line, while a swinging pendulum bob traces out a circle. These paths are known as the 'attractors' for these systems – start the system out from an arbitrary spot on the phase portrait and all paths will converge on the attractor.

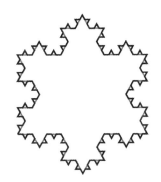

Chaotic systems typically have attractors that are fractal in nature. Fractals are geometric shapes that look the same on all length scales. A simple example is the Koch snowflake, named after Swedish mathematician Helge von Koch. Start with an equilateral triangle, then to the centre of each edge add another equilateral triangle of one-third the side length. Then repeat ad infinitum. The result is a snowflake-like pattern that you can zoom in on forever and it will just keep looking exactly the same.

Real chaotic systems are described by some quite complex fractals. Lorenz's weather systems, for example, were found to have a fractal attractor that resembled a distorted figure of eight. Much of the intimate connection between fractals and chaos was set out by the French-American mathematician Benoit Mandelbrot in his landmark 1982 book *The Fractal Geometry of Nature*.

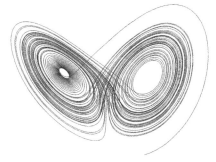

Chaos is now found everywhere scientists look. It's been seen in studies of turbulence, ocean currents, quantum theory, general relativity, engineering, biology, finance, the rotation of one of Saturn's moons, and even in psychology, where it describes the behaviour of people in groups. It seems the well-ordered world we once thought we lived in has turned out to be anything but.

The Koch snowflake (top) is a simple fractal in which each line segment dissolves into repeating equilateral triangles at diminishing scales. The Lorenz attractor (bottom) is a more complex fractal that underpins the chaos in weather systems.

The condensed idea
How unpredictability can arise from order

15 Quantum computers

Quantum computers are super-powerful calculating machines made possible by the bizarre laws of quantum physics. They are capable of performing in a few minutes tasks that would take your desktop PC longer than the age of the universe, and they're here already in laboratories around the world.

Computers today work by storing and manipulating bits of information (see page 52) using the on-off states of electrical switches called transistors. But transistors operate according to the old 'classical' laws of physics. In the early 20th century, classical physics had given way to a new view: quantum mechanics (see page 32). While classical physics remained a reasonable approximation under certain circumstances, it soon became clear that quantum theory gave the true picture of reality. In 1985, British theoretical physicist David Deutsch realized that computing as it stood was based on the wrong physics. He set about recasting the theory of computation within a quantum framework, leading to an entirely new design of computer able to vastly outperform its predecessors.

HELLO, QUBIT

In the new picture, bits of information – which can classically take the values either 0 or 1 – are replaced with quantum bits, or 'qubits', that are both 0 and 1 at the same time. This is possible because quantum mechanics

TIMELINE

1985	1994	1998
David Deutsch lays the theoretical foundations for quantum computers	Peter Shor develops a quantum algorithm for factorising large numbers	Researchers at Oxford University demonstrate the first working quantum computer

allows a quantum particle to exist in a mixture of all possible states until it's actually measured (see page 34). So if you store information in the quantum world, then it, too, will exist in a mixture of all possible states.

You might think this would be a major problem for a calculating machine, but in fact, it's the key to a quantum computer's power. When a bit of information passes through an ordinary computer processor, only the 0 or the 1 (depending on the value of the bit) gets processed. But when a qubit passes through a quantum processor, both the 0 and the 1 are processed simultaneously. Now put eight classical bits together to make a byte, and you can store any number from 0 to 255. If your quantum computer has eight qubits (a 'qubyte'), then you can store all of these numbers at the same time, and process them all in the time it takes the classical computer to crunch through just one number. In general, a quantum computer with n qubits, can store and process 2^n numbers simultaneously. Deutsch calls this 'quantum parallelism' – a nod to parallel processing on classical computers, where several processors work together on a task.

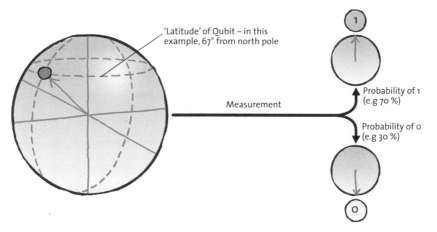

'Latitude' of Qubit – in this example, 67° from north pole

Measurement

Probability of 1 (e.g 70 %)

Probability of 0 (e.g 30 %)

Physicists represent the state of a qubit as the latitude on a sphere, where the 'north pole' represents a value of 1 and the 'south pole' represents 0. Anywhere in between is a mixture of the two, with the probability of a 1 or 0 measurement found from a trigonometric formula based on the qubit's latitude relative to the pole.

2011

The D-Wave One becomes the first commercial quantum computer

2012

1Qbit is launched as the first dedicated quantum software company

2014

Edward Snowden reveals the NSA is developing quantum computers for codebreaking

Quantum encryption

Quantum computers have major implications for national security. Modern encryption systems – used for transmitting sensitive messages securely – rely on splitting large numbers into their two factors. Sending a message just requires the number itself (which is made publicly available) but reading it requires the factors (which are fiendishly hard to calculate). Called public key encryption, it's a bit like putting a message in a padlocked box – anyone can lock it, but you need the key to open it.

Public key encryption relies on the fact that factorizing huge numbers would take a classical computer longer than the age of the universe. The bad news is that a universal quantum computer could do it in a few minutes.

The word 'parallel' in this case, however, is especially poignant. Deutsch believes in the 'many worlds' interpretation of quantum theory – whereby the strange quantum behaviour of, say, a subatomic particle is caused by interference with copies of itself in parallel universes (see page 192). In this view, a quantum computer literally derives its power from its counterparts in the universes next door. That's actually not as fanciful as it might sound: storing all of the information required to carry out some quantum computations takes more classical bits of information than there are atoms in our own universe. In Deutsch's view, quantum computers must be harnessing other universes, or else they simply wouldn't have enough memory to perform the tasks that they have already been seen to do.

COMPUTING IN ACTION

In 1998, the first functioning quantum computer was demonstrated by scientists at Oxford University, England. It only had two qubits, but it was capable of running a simple algorithm. Since then, the field has progressed considerably. In August 2015, Canadian company D-Wave Systems put its D-Wave 2X quantum computer on general sale. It has 1,024 qubits, made from superconducting loops of niobium metal. The only drawbacks are size and price: it requires a room 10 metres (33 ft) square and costs in excess of US$15 million (£10m). That hasn't stopped Google from buying one, and putting it to work training pattern-recognition algorithms that allow the Google Glass augmented reality headset to recognize objects. Aircraft developer Lockheed Martin has one too, for road-testing its flight software.

Some have criticized D-Wave's products as not being proper quantum computers, and that's true in a sense. They are not 'universal quantum computers', in that they cannot be programmed to perform any task the user requires. Instead, the D-Wave 2X uses a process called 'quantum annealing' whereby the input qubits simply evolve to their lowest energy configuration. This can be used for solving optimization problems, where the job is to find the best solution possible. Optimization has manifold applications (for example, telling a company how to spend its budget most effectively) but for large problems it can be very time-consuming. D-Wave claims that its computers can solve optimization problems 600 times faster than classical machines.

COOL IT

Building a true 'universal quantum computer' is difficult because of the fragility of quantum bits. The moment a qubit interacts with its surrounding environment, its delicate quantum state is disrupted and any quantum computations it may have been holding are lost. This is known as 'decoherence' (see page 34). Typically, a qubit will last a few seconds from creation before decoherence takes place. Researchers are trying to extend this by using cryogenic cooling techniques to reduce thermal noise, cooling the qubits to just a few thousandths of a degree above absolute zero.

Quantum computers have the potential to revolutionize industries that depend on brute force number crunching, such as finance, engineering and data analysis. They will ultimately have the power to crack today's most secure codes (see box), which has already piqued the interest of national security agencies. But one of the biggest applications will be in scientific research itself, where quantum computers will be the ultimate tool with which to simulate how quantum systems behave, deepening further our grasp on the enigmatic physics of the subatomic world.

> **QUANTUM MECHANICS IS WEIRD. I DON'T UNDERSTAND IT. JUST LIVE WITH IT. YOU DON'T HAVE TO UNDERSTAND THE NATURE OF THINGS IN ORDER TO BUILD COOL DEVICES.**
> Seth Lloyd

The condensed idea
Machines that process data in the quantum world

16 Artificial intelligence

Long before the birth of digital computers in the 1940s, scientists dreamt of building machines that can think for themselves like human beings. The applications of such artificial intelligence creations are manifold, and include everything from automated space travel to robot vacuum cleaners.

Thinking machines are a dream with a long history. Scholars of ancient civilizations, including the Greeks and Egyptians, tried (and failed) to build automata to do their bidding. It wasn't until the invention of programmable digital computers in 1943 – specifically Colossus, built by wartime codebreakers at Bletchley Park, England – that serious consideration was given to the idea of making these devices think for themselves.

ARTIFICIAL BRAIN

British mathematician Alan Turing, who oversaw the development of Colossus, had developed a theory of computation that showed how a computer could simulate any mathematical process. Therefore, if the human brain can be described using mathematics – and it seemed reasonable that it could – then a computer should be able to mimic the underlying processes. Turing conceived of a test to spot the emergence of machine intelligence (see box), and built the first chess-playing computer program. He died a few years before the world's first conference on 'thinking machines' was held in Dartmouth, New Hampshire, in 1956. The previous year, while helping organize the conference, American computer scientist John McCarthy had coined the term 'artificial intelligence'.

TIMELINE

1950	1956	1997
British mathematician Alan Turing first proposes the Turing test	The first artificial intelligence conference takes place in Dartmouth, New Hampshire	IBM's chess computer Deep Blue beats grandmaster Garry Kasparov

The climate of optimism into which the field was born was soon tempered in the 1960s, as researchers realized that breathing human traits into a machine was going to be much harder than they had at first imagined. By the mid-1970s, most government funding for AI research had been withdrawn and the field was stagnating.

NEATS AND SCRUFFIES

Research into AI at this time broke down broadly into two camps, dubbed 'neat' and 'scruffy'. The neat approach utilized standard techniques from computer programming to try and engineer behaviour that resembled intelligence. On the other hand, the scruffy approach was more organic – even referred to as 'anti-logic'

The Turing test

British computer scientist and codebreaker Alan Turing suggested a test in 1950 to determine whether or not a computer could be considered intelligent. The basic idea was that a human judge would simultaneously hold conversations with a computer and a real human, and if the judge was unable to tell which was which, then the computer – to all intents and purposes – could be considered intelligent.

The test led to the development of 'chatbots', AI software entities developed for the sole purpose of conducting text chat sessions with people – to try and pass Turing's test. The first chatbot, called ELIZA, was developed in 1966 by German-born computer scientist Joseph Weizenbaum. In 2014, a chatbot called 'Eugene Goostman', developed by a team of programmers from Russia, was the first to actually pass a Turing test, in an event held at the University of Reading.

Today, chatbots are used in online customer service, while the Turing test is used in the CAPTCHA (Completely Automated Public Turing test to tell Computers and Humans Apart) system commonly used for online identification of human customers. The Turing test also became known as 'The Imitation Game', which was later the title of a 2014 movie about Alan Turing's life and work.

by some. It amounted to bolting together bits of hardware and software, to see what kind of behaviours could be cooked up.

Neural networks were one major technology to emerge from the scruffy approach. These are software systems that emulate groups of brain cells

2004
NASA's Mars Exploration Rovers autonomously navigate around the Red Planet

2009
Google launches 'driverless' cars that can run without human input

2014
Eugene Goostman becomes the first chatbot to pass a Turing test

and the connections between them. Like brain cells, they can learn from experience – helping a computer to make intelligent decisions. Neural nets are now just one of a number of 'machine learning' techniques that are used in everything from detecting fraud to recommending songs you might enjoy hearing on your smartphone.

> **I'M SORRY, DAVE. I'M AFRAID I CAN'T DO THAT.**
> HAL 9000 computer,
> *2001: A Space Odyssey*

The development of 'expert systems' in the 1980s gave artificial intelligence a much-needed boost. These are AIs that attempt to replicate the knowledge of human specialists in a particular field – for example, making rapid and accurate decisions in a crisis, or diagnosing infectious diseases. The recovery was short-lived, however, and by the late 1980s the field was languishing once more. It was another decade until, at the turn of the millennium, artificial intelligence saw a more decisive change in its fortunes. Vastly more powerful computers and better communication between AI's disparate subfields, spanning disciplines from cybernetics to linguistics, drove research forwards at a new pace.

GAME THEORY

In 1997, a chess computer called Deep Blue outsmarted Russian grandmaster Garry Kasparov. Deep Blue drew three of the six games played, won two and lost one, winning the overall match. The late 1990s also saw breakthroughs in AI toys, with the release of the Furby and Sony's AIBO robot dog. In 2002, the Roomba, an artificially intelligent robot vacuum cleaner, went on sale. Meanwhile, NASA landed artificially intelligent robots, the autonomous Spirit and Opportunity Exploration Rovers, on Mars during 2004.

In 2009, internet giant Google developed the world's first working driverless car. And in 2010, the company released a version of its Android mobile operating system that supports voice commands, using AI software to spot words and phrases. Recently, game-playing AIs have not only excelled at chess, but also proved themselves capable of beating human players at poker and checkers. In 2011, a machine called Watson, developed by IBM, trounced two former champions of the US TV quiz show *Jeopardy!*. And in 2014, a Russian-built chatbot called Eugene Goostman became the first machine to pass a Turing test (see box on page 65).

SUPERINTELLIGENCE

One reason for the recent upsurge in AI breakthroughs is the ever improving power of computers. The power of modern PCs has been shown to roughly double every 18 months – a progression known as Moore's law, after the Intel founder Gordon Moore who first noticed it. This trend, and the improvements it has brought to artificial intelligence, has led some to speculate that we could one day create an AI 'superintelligence'.

Such an entity would be far smarter, and able to redesign and improve itself far quicker, than its human creators. This would create an even smarter superintelligence. The process repeats until the AI becomes infinitely smart, which is a state known as the 'singularity' – a term put forward in 1958 by the Hungarian-born mathematician John von Neumann. What might such an entity want to do with us, its inferior creators?

Computer scientist Ray Kurzweil has stated that we will reach the singularity by the end of this century. Some critics argue, however, that true AI may be beyond even infinite technological advancement. It could be that the increasing complexity of technology will serve as a natural brake to progress, making it impossible for a computer to replicate. Or maybe human intelligence can't be described by mathematics after all. Others are sceptical, perhaps naively, that humanity could ever blindly engineer its own demise.

The condensed idea
Machines that can think

17 Atoms and molecules

Everything around us is composed of atoms and molecules – the tiny particles that make up our DNA, the air that we breathe, the food that we eat and the pages of this book. They are among the wonders of the natural world, and their discovery marked a new dawn in science.

It is said that there are more atoms in a cup of water than there are cups of water in all the oceans on Earth. This dazzling statement emphasizes the minuscule size of the atom: each is about one-tenth of a millionth of a millimetre across. Just one cell in the human body can contain 100 trillion atoms at any one time. Yet bizarrely, each atom is mostly empty space with its mass being concentrated in the central nucleus.

Mankind has been fascinated by the structure of matter since the time of the Ancient Greeks. Leucippus of Miletus is believed to be the first person to propose atomic philosophy, followed by his more famous fifth-century CE disciple Democritus of Abdera. They believed that everything is composed of atoms, and that they are infinite in number and indestructible (the name 'atom' comes from the Greek word *atomos*, meaning indivisible).

Democritus also believed that atoms gave matter its unique characteristics. For instance, he proposed that iron atoms were strong and solid whereas

TIMELINE

400 BCE	1808	1811
Democritus proposes that all matter is constructed of solid, indestructible units	English chemist John Dalton publishes his ground-breaking atomic theory	Italian scientist Amedeo Avogadro proposes that molecules are formed by joining atoms

ice had smooth, spherical atoms. It was a fanciful theory but at its heart Democritus was right. Deconstruct all matter in the universe and eventually it can be broken down into the tiniest part of any given element, the atom, and the atom's structure determines the element's chemical properties.

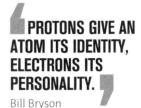

PROTONS GIVE AN ATOM ITS IDENTITY, ELECTRONS ITS PERSONALITY.

Bill Bryson

DALTON'S THEORY

Modern atomic theory began to evolve in 1805, when English chemist John Dalton put forward the notion that matter is made up of indivisible chunks that he, too, called atoms. He had noticed a result obtained by the French chemist Joseph-Louis Proust – that substances tended to combine during chemical reactions in fixed, whole-number ratios. For example, when oxygen combines with tin, the amount of oxygen that reacts is always a fixed multiple of 13.5 per cent of the mass of tin, depending on the type of reaction. The implication is that an atom of oxygen weighs 13.5 per cent the mass of an atom of tin.

On the basis of these findings, Dalton established his own atomic theory, which he specified using five axioms:

1. All chemical elements are made of tiny particles called atoms.
2. Atoms of a particular element are identical in size, mass and other properties.
3. Atoms of a given element differ from those of other elements.
4. Atoms are indivisible and indestructible.
5. Compounds can be formed when atoms of one element combine with those of another element and they do so in equal quantities.

Science has since disproved some of Dalton's axioms, but the first and third still hold true and are the most important. He first presented his theory to the Manchester Literary and Philosophical Society in 1803, and expanded on it in

1905
Einstein proves the existence of atoms by studying pollen grains

1909
Jean Baptiste Perrin makes the first estimate of Avogadro's constant

1926
Perrin wins the Nobel Prize in Physics for proving the existence of molecules

Chemical bonds

How do atoms fuse together to create molecules? Within the atom, the nucleus has a positive charge and is surrounded by a cloud of negatively charged electrons. As atoms approach each other the electron clouds interact to bond and form a molecule.

There are two types of chemical bond. The covalent bond, where two atoms share a pair of electrons, is found in non-metallic elements, such as carbon, oxygen and hydrogen. In the ionic bond, atoms gain or lose electrons, thus creating charged particles called ions. The atoms of non-metals gain electrons to acquire a negative charge, while hydrogen and metal atoms lose electrons, creating a positive charge. The opposite charges create the force of attraction that holds the molecules together.

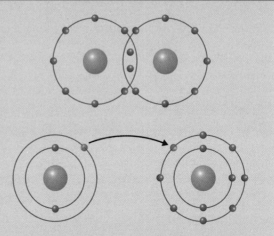

Atoms in a covalent bond (top) share electrons, while an ionic bond (bottom), involves an electron being transferred so that one atom gains a negative charge and the other is left with a positive charge.

his 1808 book *A New System of Chemical Philosophy*. Dalton paved the way for significant developments in the study of chemistry, but it was almost a century before Albert Einstein finally proved the existence of atoms by interpreting the zigzag motion of pollen grains, observed under a microscope. He deduced that they were being buffeted by collisions with unseen atoms.

CHEMICAL COMPOUNDS

In 1811, a little-known Italian chemist called Amedeo Avogadro built on Dalton's theory, proposing that atoms could bond together following chemical reactions to create molecules. The word 'molecule' was originally put forward by 17th-century French philosopher René Descartes, but the term had been used interchangeably with atoms. Avogadro set the record straight by declaring that atoms are indivisible units of chemical elements, while molecules are the fundamental units of chemical compounds, formed by locking atoms together. Avogadro's seminal work also included the first formal statement of what would become known as Avogadro's law. This states that equal volumes of all gases, at the same temperature and pressure, have the same number of molecules regardless of their physical properties and chemical nature. Strictly speaking, the law only applies to an 'ideal gas' where none of the molecules exert force on each other, but it is still a good approximation in most cases.

AVOGADRO'S CONSTANT

The Italian chemist is also connected with another key breakthrough regarding the number of atoms in a material. He showed that if you add up the masses of all the atoms in a particular molecule, you arrive at a quantity called the 'molecular mass'. For instance, carbon has an atomic mass of 12 and oxygen has an atomic mass of 16. Therefore a molecule of carbon dioxide (CO_2) has a molecular mass of 44 (one carbon plus two oxygen). Therefore, if you have a quantity of a substance in grams that is equal to the substance's molecular mass, then the number of molecules will always be the same.

Today, this number, known as Avogadro's constant, is known to have the value 6.022×10^{23} (that's roughly a 6 followed by 23 zeros). This number of molecules is often referred to by scientists as the 'mole' of a substance: for instance, 44 grams (1.55 oz) of CO_2 contains one mole of molecules. Avogadro's constant was not defined by the Italian himself, but by French scientist Jean Baptiste Perrin. He suggested that the value be named after Avogadro to honour the Italian's key discoveries in the field of molecular science (work that had gone largely unrecognized in his own lifetime). Perrin went on to prove the existence of molecules and won the 1926 Nobel Prize in Physics.

The condensed idea
Matter is made up of tiny particles

18 The periodic table

In the 1860s, Russian chemist Dmitri Mendeleev transformed our understanding of the chemical elements with a ground-breaking discovery. Fascinated by the elements, Mendeleev believed they could be ordered according to their properties. The result was the periodic table, a unique and elegant construct of the chemical building blocks of our universe.

Every school chemistry lab has a periodic table on its walls – a chart of letters, numbers and colours, ordering the chemical elements from which our universe is built. Pupils tend to be either transfixed or bewildered by the complex information and diversity of the elements contained within.

Elements are the fundamental building blocks of the world around us. They can exist in their pure form as atoms – as opposed to compounds, where the fundamental units are molecules formed by atoms joining together (see page 68). By the early 1860s, there were 62 known elements and for years scientists had tried to discover patterns among them. It was Dmitri Mendeleev who would finally succeed.

ORDER FROM CHAOS

Mendeleev started by listing each element's atomic mass (now known to indicate the total number of neutrons and protons in the nucleus, though these subatomic particles were not discovered until the 20th century). He wrote down each element on a piece of card, with its atomic mass and a few physical

TIMELINE

1789	1862	1864
French scientist Antoine Lavoisier groups the elements by their basic properties	French geologist de Chancourtois discovers how to calculate atomic masses of elements	Scottish chemist John Newlands discovers the beginnings of periodicity between elements

properties. Initially, he organized his table as a line of elements in order of increasing mass. But he then discovered that if he divided this sequence into strips and arranged the strips into rows, as a table, the elements in each column tended to share similar properties. For example, Mendeleev's left-most column listed sodium, lithium and potassium, all of which are metals that react violently when dropped into water. The columns of elements in the table were referred to as 'groups', due to their similarities, while the term 'period' was applied to the rows, due to the repetition of properties.

Mendeleev published the first iteration of his 'periodic table' in 1869, but it was evolving all the time. If an element seemed to be in the wrong place, Mendeleev would move it to a position where it fitted the pattern he had created. He left gaps in the table where he believed undiscovered elements would later slot in, and successfully predicted the properties of five such elements and their compounds.

> **I WAS GIVEN A BOOK ABOUT THE PERIODIC TABLE OF THE ELEMENTS. FOR THE FIRST TIME I SAW THE ELEGANCE OF SCIENTIFIC THEORY AND ITS PREDICTIVE POWER.**
>
> Sidney Altman

For instance, he inferred that there was an unfilled space one down from aluminium and estimated that the 'missing' element would have an atomic mass of 68. He dubbed the element eka-aluminium, and was proved right six years later when the French chemist Paul-Émile Lecoq de Boisbaudran isolated the element gallium, which matched all the predicted properties of eka-aluminium. Scandium and germanium were discovered in the 1879 and 1886 respectively, further reinforcing the validity of the table. When Sir William Ramsay discovered the inert gases in the 1890s, these created an extra group on the table.

ATOMIC CHARGE

The periodic table has since expanded to feature 118 elements. It was Mendeleev's flexible approach, reordering the table when he deemed it necessary, that allowed the true principle underpinning how the elements are arranged to emerge: not by their mass, but by their 'atomic number'.

1869	1875	1890s
Mendeleev makes history by organizing the 62 known elements into the periodic table	The element gallium is discovered by de Boisbaudran, as predicted by Mendeleev	William Ramsay discovers a new group of the table, the inert gases

Dmitri Ivanovich Mendeleev (1834–1907)

Mendeleev was born in Tobolsk, Siberia, one of at least 13 children. The family fell on hard times when Mendeleev's father developed blindness while Dmitri was still a boy. His mother, Mariya, was determined that her son would receive an education and in 1849 they embarked on the cross-country journey to St Petersburg, where he was accepted at the Central Pedagogic Institute to read physics and mathematics. He graduated with a gold medal for excellence and continued his studies in St Petersburg and the University of Heidelberg.

Mendeleev was appointed professor of general chemistry at the University of St Petersburg in 1867 and taught there until 1890. He wrote a large number of articles and books, including his landmark title *Osnovy khimii* (*The Principles of Chemistry*), which was translated into many languages.

Mendeleev's great discovery came in February 1869. He had planned to visit a cheese factory that day but the weather was bad so he stayed at home and worked instead. Initially, the periodic table did not receive much attention from chemists. However, the discovery of elements that he predicted, such as gallium, led to the table becoming one of the great mainstays of theoretical chemistry.

The question remains, why didn't the 'father of the periodic table' win a Nobel Prize? Mendeleev was nominated, in 1905 and 1906, but one of the judges considered his achievements to be too old and well known. However, by the time of his death in 1907, Mendeleev had garnered international recognition and had won many other awards for his findings.

Most of us recognize elements by their chemical symbol (such as H for hydrogen), but they can also be classified by atomic number. Whereas the atomic mass reflects the combined number of protons and neutrons in the nucleus, the atomic number is determined solely by the number of protons. Hydrogen, with only one proton, has atomic number 1.

Protons each carry a single unit of positive electric charge, so atomic number is a measure of the charge of the nucleus. The connection between atomic number and nuclear charge was proved in 1913 by Henry Moseley, a British physicist, and led to the discovery of further elements. What hadn't been established by this point was exactly why elements in the same group share chemical properties. An emerging branch of science called quantum theory (see page 40) was soon to provide the answer.

QUANTUM CHEMISTRY

The atomic nucleus is surrounded by a cloud of negatively charged electron particles. Quantum theory dictates that these electrons are organized into concentric 'shells' around the nucleus. Elements in the same group as one another tend to have the same or similar numbers of electrons in their outermost shell, and these are the electrons that are most readily available to undergo chemical reactions with other atoms – hence, elements within groups share similar chemical properties regardless of how many additional layers of electrons underlie the outermost shell.

Each element has a fixed number of protons in its nucleus, but the number of neutrons can differ, leading to 'isotopes', atoms that have the same chemistry but behave differently in nuclear reactions (see page 44). What's more, atoms in bulk can be arranged differently within a sample of an element, leading to 'allotropes' with different physical properties. For instance, we have O_2, which has two oxygen atoms bonded together, and O_3 or ozone, with three linked oxygen atoms.

Over time, Mendeleev's original table of eight groups has grown, with the discovery of new elements. More recent additions include the lanthanides, actinides and elements with unknown chemical properties that can only be synthesized a few atoms at a time in a laboratory. Mendeleev was never awarded a Nobel Prize for his work, but fittingly he now has the rarer honour of having an element named after him, number 101 mendelevium.

The condensed idea
The hidden order underlying the chemical elements

19 Radioactivity

Some chemical elements are inherently unstable. Their atoms slowly break apart over time to give out a steady stream of particle fragments and energy, known collectively as radioactivity. Some radioactive materials are hazardous to health, but others can, with careful handling, be put to use in applications such as medical diagnosis and cancer therapy.

Radioactivity was discovered by Henri Becquerel in 1896. The French physicist was fascinated by X-rays, which had been discovered the previous year by the German physicist Wilhelm Röntgen. Becquerel wondered whether X-rays had any connection with phosphorescence (whereby some materials slowly re-emit absorbed sunlight) and began testing his hypothesis. Röntgen had observed that X-rays could fog sealed photographic plates and Becquerel tried a similar tack. He wrapped photographic plates in black paper and placed phosphorescent salts on top of them. During his experiments, Becquerel discovered that only uranium salts (potassium uranyl sulfate) were able to fog the sealed plate.

At first, Becquerel believed that the uranium had absorbed energy from the Sun and emitted it as X-rays, but an overcast day in Paris proved this hypothesis wrong – the fogging still appeared on the plate. Becquerel deduced that some invisible radiation had passed from the uranium through the black paper to cause the fog. The phenomenon was later named 'radioactivity' by Marie Curie who, with her husband Pierre, discovered the highly radioactive elements radium and polonium.

TIMELINE

1896	1898	1899
Henri Becquerel observes natural radiation produced by uranium	French scientists Marie and Pierre Curie coin the term 'radioactivity'	Ernest Rutherford observes that uranium radiation has alpha and beta particles

RADIOACTIVE EMISSIONS

Continued study of radioactivity by Becquerel and other scientists revealed that the radiation takes three forms – alpha, beta and gamma. Alpha radiation is made up of heavy, short-range particles. Uranium, radium and thorium are all alpha-emitters and occur naturally in the environment, in rocks, soil and water. Alpha particles travel slowly, losing their energy rapidly in air, and cannot pass through human skin or clothing.

Beta radiation is made up of lighter particles that can travel several metres in air. It can penetrate the germinal layer of skin, where new cells are being produced. Strontium-90 is a beta-emitter, and large areas of Russia and western Europe were contaminated with this radioactive isotope following the Chernobyl disaster in 1986. Gamma radiation, meanwhile, is electromagnetic energy, with no limit to its reach. It can penetrate deep into human skin, hence its use in medicine to destroy cancer cells. But this also makes it potentially lethal.

From the late 1890s, the work of New Zealand-born physicist Ernest Rutherford shed light on the causes of radioactivity. Working in Montreal, he and the chemist Frederick Soddy devised the disintegration theory, essentially arguing

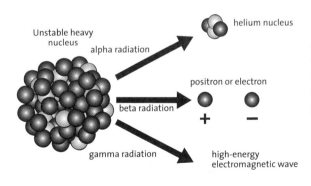

Unstable heavy nucleus

alpha radiation

helium nucleus

beta radiation

positron or electron

gamma radiation

high-energy electromagnetic wave

Radioactivity generates three forms of emission – alpha particles, beta particles and gamma rays.

1903
Rutherford and Soddy put forward the atomic disintegration theory of radioactivity

1913
Hans Geiger reveals his iconic radiation detector, the Geiger counter

1986
Chernobyl nuclear power planet explosion releases 5,200,000,000,000,000,000 Becquerels of radioactivity

that the radiation is caused by atoms breaking apart into their constituent particles. However, Rutherford's further experiments into the scattering of alpha particles led him to discover that most of the mass of an atom, including all the neutrons and protons, is centred in its nucleus (see page 72).

Armed with this knowledge, it soon became clear that radioactivity wasn't just about the break-up of an atom, but the disintegration of the very nucleus itself. Alpha particles were revealed to be nuclei of helium, each made up of two protons and two neutrons, formed when this quartet of particles are ejected from an atomic nucleus. Beta particles, meanwhile, were nothing more than electrons. These particles don't normally exist inside the nuclei of atoms, but it was shown that under certain circumstances, an electrically uncharged neutron particle can decay into a positively charged proton that stays in the nucleus, and a negatively charged electron, which is ejected. Gamma radiation eventually proved to be nothing but high-energy electromagnetic waves (see page 16), released when a nucleus drops from a higher- to a lower-energy state.

ACTIVE ELEMENTS

Not all chemical elements are radioactive: generally speaking, it is a feature of elements with heavy atomic nuclei – those with more than 83 protons (uranium, for instance, has 92). These nuclei find it harder to hold together because electrostatic repulsion between all those protons (which all have the same positive electric charge) tends to push them apart – and this makes such heavy nuclei unstable.

Scientists quantify the degree of radioactivity of a nucleus using the term 'half-life'. This is basically the amount of time it takes for half of the nuclei in a sample to decay. Radioactive decay is essentially a statistical process: no law of physics can tell you when a specific atom will decay. But we can say, given a sample containing a large number of atoms, what the average rate of decay will be, and this is what the half-life tells you. Half-lives can be enormous: for example, plutonium-239 (plutonium with a total of 239 particles in its nucleus, one of the waste products from nuclear power plants) has a half-life of 24,000 years.

FRIEND AND FOE

Radioactivity has found many uses. The process is used in radiocarbon dating where a radioactive form of carbon with a known half-life (5,730 years) can indicate the age of materials. Ancient artefacts, such as the Dead Sea Scrolls and archaeologial remains, have been dated by this procedure.

More benefits came in the field of nuclear medicine. Radioactive tracers can be used to warn of abnormal functioning of an organ in the body. Radioisotopes with short half-lives such as technetium-99 are used to ensure the radiation does not stay in the body for a long time. Meanwhile, almost two-thirds of cancer patients receive radiation therapy during their illness. Radiation is also used to sterilize medical instruments: they are sealed in an airtight bag, then bombarded with gamma rays that can penetrate the bag. This kills the bacteria that may lurk even on 'clean' instruments, so they can be kept sterile until the bag is opened for use.

Radiation hazard

In the early days following the discovery of radioactivity, scientists who became transfixed by its properties took massive risks. Pierre Curie voluntarily exposed his arm to radium for a few hours, which resulted in a lesion that took several months to heal. Becquerel suffered a burn after carrying a glass tube containing radium salt in his vest pocket. He later commented of radium that 'I love it, but I owe it a grudge'.

Marie and Pierre Curie took few precautions in handling radioactive matter and in 1934 Marie died of aplastic anaemia, a condition that prevents bone marrow from producing sufficient blood cells. The Curie's papers and Marie's cookbooks are still too radioactive to handle today.

Yet radioactivity is also a dangerous phenomenon, deserving of our respect. A stark reminder of this came in 1986, when the Ukrainian nuclear power plant at Chernobyl exploded, scattering radioactive debris over a wide area. It is estimated that the level of radioactive contamination will render Chernobyl uninhabitable for the next 20,000 years.

The condensed idea
Spontaneous disintegration of atomic nuclei

20 Semiconductors

Semiconductors have revolutionized the way we live, work and communicate. They led to the miniaturization of electronics, which gave us the microchip, and are present in almost every electronic device we use today, from radios and televisions to personal computers, satellite navigation and the mobile phone.

A semiconductor material is one that conducts electrical current, but only partly. It does not conduct sufficiently well enough to be called a conductor, such as copper, and neither does it insulate enough to qualify as an insulator, like glass. Most semiconductors are crystals, made up of materials such as silicon and germanium. Their usefulness stems from the fact that their conductivity can be controlled precisely using an electric current. This enables engineers to create devices that can amplify or switch electrical signals – the latter especially are integral to so many of the devices we rely on from day to day.

THE BIG BREAKTHROUGH

The first practical semiconductor device was created at the Bell research laboratory of the American Telephone and Telegraph Company in New Jersey. In 1945, a research team was established to develop semiconductor amplifiers, led by the British-born physicist William Shockley. Shockley preferred to work from home, which allowed his colleagues John Bardeen and Walter Brattain to continue their research independently. In 1947, Bardeen and Brattain created a very rough amplifier consisting of two gold contacts, held together by a plastic wedge, with the thin end of the wedge resting on a piece of semiconducting

TIMELINE

1940	1947	1951
Ohl discovers the 'P-N' junction, a key element of semi-conductor devices	Shockley, Bardeen and Brattain invent the first working semiconductor device, the transistor	William Shockley invents the sophisticated bipolar junction transistor

germanium. The device worked: a large current passing through one of the gold contacts and the germanium could be controlled with the output from the other contact, essentially amplifying the signal.

Bardeen and Brattain had created the first 'transistor' – a portmanteau of 'transfer' and 'resistor' (a device for regulating electric current). Bell Labs announced the discovery in 1948. Shockley was thrilled at the development, but deeply annoyed that the duo had forged ahead without him. Consumed by professional jealousy, he shut himself away and attempted to refine the device alone.

N-TYPE AND P-TYPE

Different kinds of semiconductors can be made by adding impurities, or 'doping' as it is known. An 'n-type' semiconductor is made by adding an element with an excess of electrons – for instance, phosphorus is a common dopant for silicon. A 'p-type' semiconductor, on the other hand, is created by adding a dopant that is deficient in electrons, such as boron. As a result, an n-type semiconductor has excess electrons, while a p-type semiconductor can be thought of as having positively charged 'holes', left behind by the missing electrons.

John Bardeen (1908–91)

John Bardeen is the only person to win the Nobel Prize in Physics twice: in 1956, for inventing the transistor with Brattain and Shockley; and in 1972 for work on the theory of superconductivity (see page 84) with Leon N. Cooper and John Robert Schrieffer.

Bardeen was born in Wisconsin in 1908, the son of the Dean of the University of Wisconsin Medical School. He studied for a degree in electrical engineering and was awarded a doctorate in mathematical physics from Princeton in 1936.

He began his career at the Bell research laboratory, New Jersey, in 1945, where he collaborated with Brattain to invent the transistor. He left in 1951 to become professor of physics at the University of Illinois. Bardeen was a quiet, unassuming man who shunned the limelight. Nevertheless, in 1990, he was among *Life* Magazine's '100 Most Influential Americans of the Century'.

In 1951, Shockley came up with an improved version of Bardeen and Brattain's transistor. It became known as a 'bipolar junction transistor' and consisted of three alternating layers of n-type and p-type semiconductors. Two different

1956
Bardeen, Brattain and Shockley win the Nobel Prize in Physics

1958
The semiconductor microchip is invented, paving the way for digital electronics

2015
IBM announces the smallest ever transistor – at 7 nanometres in size

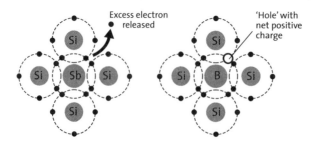

Excess electron released

'Hole' with net positive charge

Silicon (Si) is doped with antimony (Sb) to give an n-type negatively charged semiconductor (left), or boron (B) to produce a p-type positively charged semiconductor (right).

THE ABILITY TO USE SEMICONDUCTOR SEQUENCING TO PROVIDE A MEDICAL DIAGNOSIS IN JUST A FEW HOURS THAT ONCE TOOK A FEW DAYS IS A CRUCIAL STEP IN SAVING THE LIVES OF PATIENTS.

Professor Christopher Toumazou

configurations of the device are possible. In an NPN transistor, increasing the voltage to the middle p layer boosts the number of electrons within it, increasing the conductivity across the two outer layers and in turn increasing the current. In a PNP transistor, the process is much the same, though governed not by the electrons in the middle layer but by the number of positive holes.

The device basically allowed a small electric current applied to the middle layer to control a much larger current flowing between the outer two layers. It was a vast improvement on the bulky, inefficient vacuum tubes previously used in electronics. Bardeen, Brattain and Shockley were jointly awarded the 1956 Nobel Prize in Physics for their work on semiconductors and the transistor.

FROM RADIOS TO MICROELECTRONICS

Transistors first went on sale in 1949, but few people realized their significance at the time. The transistor radio was launched in 1954 and, after modifications that cut retail price, it became hugely popular, with billions being sold in the 1960s and 1970s. But there were far more significant applications to come.

Shockley, Bardeen and Brattain's transistor could be used not only as an amplifier but also as a switch, flicking tiny amounts of electricity on and off. The on/off state of a single switch can store the state of a binary digit, or 'bit' of information (see page 52). Groups of semiconductor switches form logic gates, the building blocks of digital circuits, which perform functions and process information. Logic gates are the basis of digital electronics and underpin the workings of your personal computer, digital camera, television and mobile phone. Today's transistors can turn themselves on and off 3 billion times per second, processing information at lightning speeds.

Without transistors, computers would still be built from vacuum tubes – glass tubes containing electrodes with a gap between them to regulate the flow of electrons. These took up a large amount of space, which is why early computers occupied an area the size of a room. The first transistor computer was built in 1953 at the University of Manchester. Two years later, IBM announced the launch of its first transistor calculator, the IBM 608.

In 1958, American engineers Jack Kilby and Robert Noyce made history when they invented the semiconductor microchip, or silicon chip, which miniaturized the technology, packing many transistors onto a small piece of semiconductor, and led to the development of the personal computer. The first transistor was about 1.3 centimetres (0.5 in) across. Today, they are so small that a single microchip can hold billions of them. Semiconductors are now a core component of every electrical device that we use and have made the technology available to the masses. They have revolutionized industry, communication, medicine and space exploration. Their invention marked the dawn of the digital era.

Tiny transistors

Shrinking the size of transistors has led to cheaper and faster electronic devices. The physical limit on transistor size is 5 nanometres (nm) – any smaller and you enter the realm of quantum physics, where matter and energy work in mysterious ways. One such effect is electron tunnelling. If the material is 5 nm thick (or less), electrons will tunnel through it. Transistors are designed to control the flow of electrons, so this becomes a problem. In 2015, IBM announced that it had made a 'test' chip with a transistor just 7 nm in size. To give an idea of scale, a human hair is 80,000nm thick. At the time of writing, the most transistors on a commercially available chip is 5.5 billion, in the Intel Haswell-Xeon-EP.

The condensed idea
Making microchips

21 Superconductors

At low temperatures, the electrical resistance of certain materials suddenly vanishes – a phenomenon known as superconductivity. Superconductors have already found applications in transportation, super-efficient generators, medical imaging devices and particle accelerators. The race is now on to make them work at room temperature.

A superconductor is a material that can conduct electricity with zero resistance. This has great benefits because resistance causes energy in an electrical current to be lost as heat. Superconductors offer the potential to radically improve electrical power technology, through minimizing such waste. Currently, as electricity is generated, 10 to 15 per cent is dissipated as heat from transmission cables. Use of superconductors could cut this to zero, with a huge impact on energy efficiency and the environment.

The phenomenon was discovered by Dutch physicist Heike Kamerlingh Onnes, professor of experimental physics at the University of Leiden. In 1904, he established a large cryogenics laboratory to study the behaviour of materials at very low temperatures. At the time, a number of scientists, including the Irish physicist Lord Kelvin, believed that at such low temperatures the electrons carrying the current would freeze, increasing resistance, but Onnes believed that resistance would fall gradually. In 1911, while studying conductivity in a wire made out of mercury, he noticed that if he dropped the temperature to 4.2 degrees above absolute zero (-269°C or -452°F), the resistance of the mercury suddenly disappeared. It was a

TIMELINE

1911	1933	1957
Onnes discovers superconductivity after applying current to mercury near absolute zero	Meissner and Ochsenfeld discover that superconductors expel any magnetic field applied to them	Bardeen, Cooper and Schrieffer publish the BCS theory of superconductivity

momentous discovery, and Onnes was awarded the Nobel Prize in Physics in 1913 for his breakthrough.

Scientists were fascinated but perplexed by the phenomenon. No one understood it. In 1933, the German scientists Walther Meissner and Robert Ochsenfeld noted that superconductors expel any magnetic field within them. The Meissner effect, as it is known, is so powerful that if a magnet is placed on a superconductor it will levitate and hover above it.

> THUS THE MERCURY AT 4.2K HAS ENTERED A NEW STATE, WHICH, OWING TO ITS PARTICULAR ELECTRICAL PROPERTIES, CAN BE CALLED THE STATE OF SUPERCONDUCTIVITY.
> Heike Kamerlingh Onnes

RESISTANCE IS FUTILE

In 1957, the American scientists John Bardeen, Leon Cooper and Robert Schrieffer, at the University of Illinois, figured out what was happening. In a normal conductor such as copper, free electrons slosh around in between the atoms in the material. When a voltage is applied, it's the flow of these electrons from negative to positive that makes an electric current. However, electrons aren't the only thing in the material – there are also positively charged atomic nuclei. Heat makes the lattice of nuclei in the material vibrate and collide with the electrons, impeding the flow of current – this is the cause of electrical resistance. Cooling lowers the resistance by reducing the vibrations.

In a superconductor, resistance vanishes entirely. The BCS theory (taken from the initial letters of the researchers' surnames) supposed that this happens because electrons lock together in 'Cooper pairs' that slip through the lattice of atomic nuclei without collision. Broadly speaking, the negative charge of the first electron in the pair attracts the positive atoms, distorting the lattice inwards. This distortion creates a concentration of positive charge that pulls the following electron forward, and keeps the current going.

1962	1986	2006
The first commercial superconducting wire is developed by firm Westinghouse	Johannes Georg Bednorz and Alex Müller create the first high-temperature superconductor	The largest-ever superconducting magnet – for the Large Hadron Collider – is switched on

In superconductors at low temperatures, one electron in a 'Cooper pair' pulls positive atomic nuclei together, creating a positively charged region that pulls the other electron towards it.

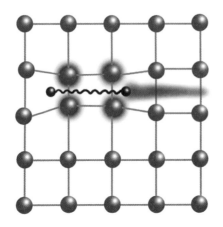

GETTING WARMER

The problem with superconductors was that they required extremely low temperatures, which meant using expensive cooling fluids like liquid helium (see box). A breakthrough came in 1986, when IBM scientists Johannes Georg Bednorz and Karl Alexander Müller discovered a copper-ceramic blend that would superconduct at 30 kelvin (K) (-243°C or -405°F) – still very cold, but warmer than Onnes's mercury wires. The duo won the Nobel Prize in Physics for their achievement. Materials have since been created that superconduct at 127 K (-146°C or -231°F). These materials became known as 'high-temperature superconductors' and their discovery led to exciting new developments in electronics, medicine, power and travel.

For example, superconductors are the driving force behind Japan's 'bullet train', which hovers using magnetic levitation. In 2015, the seven-car maglev broke the high-speed rail world record with a top speed of 600 kilometres per hour (373 mph). Thanks to the Meissner effect, its superconducting magnets levitate it off the ground, so that it floats 10 centimetres (4 in) above special guideways.

Particle accelerators, such as the Large Hadron Collider at CERN in Switzerland, also make use of superconductivity. They involve long underground tunnels where particles are boosted to near light speed, before being smashed together to create a shower of debris that reveals even smaller, fundamental subatomic particles. CERN's 27-kilometre (17-mile) circular tunnel features over 1,500 superconducting magnets to create the magnetic field that guides the particles on their way. Each is 15 metres (50 ft) long, weighs 35 tonnes and generates a magnetic field 100,000 times more powerful than the Earth's.

MEDICAL MAGNETS

Superconductors are a key component in magnetic resonance imaging (MRI) machines that are used to diagnose illnesses such as heart disease and cancer. MRI systems use a powerful magnetic field to stimulate electromagnetic emission from water molecules in the body. The emission can then be measured and used to form a diagnostic image. The field is created by a superconducting magnet, consisting of wire coils through which current is passed. The magnet is kept at a temperature of -269.1°C (-452.4°F) by a bath of liquid helium.

Meanwhile, superconducting electricity generators have the power to change the world. Standard generators use ordinary copper wire, coils of which are mechanically rotated inside a magnetic field to produce a current. Generators that replace the copper wire with superconducting wire are almost 100 per cent efficient and around half the size of conventional models. Physicists in Finland have calculated that if all European Union countries used such generators in their power plants, annual carbon emissions could be reduced by 53 million tonnes. Even so, energy is still lost in the transmission of electricity from electrical generator to end user. The only way this will be solved is by finding a superconductor that can operate at room temperature – and that is the next big challenge.

The coldest place on Earth

In preparing for his experiments on low-temperature conductors, Heike Kamerlingh Onnes made his home town of Leiden the coldest place on the planet. To keep conductors at the low temperatures required he needed a coolant, and on 10 July 1908, he found it. As he cooled helium gas close to absolute zero, it switched to a liquefied state – the first time liquid helium had ever been made. Onnes continued to cool the helium to just 0.9°C (1.6°F) above absolute zero. It is the most effective refrigerant known.

The condensed idea
Opening up electricity's fast lane

22 Buckyballs and nanotubes

Before 1985, scientists believed there was nothing new to discover about carbon. Then came the discovery of fullerenes, unique forms of carbon that have huge technical potential. They are the basis of materials 20 times stronger than steel, and offer potential medications for AIDS and cancer.

Carbon is one of the key molecules of life. It is found in almost every cell of the human body and is the fourth most abundant element in the Milky Way. For many years we have known of three main carbon allotropes (different molecular arrangements of atoms). These were soot, graphite and diamond. In 1985, an international team of scientists made a landmark discovery, revealing a new allotrope of carbon that would offer exciting potential in the fields of engineering, medicine and technology.

The story begins at the University of Sussex where chemist Harold Kroto was interested in examining the properties of carbon molecules that cluster in the atmospheres of red giant stars. He used a technique called microwave spectroscopy to analyse them. Then, during a visit to Rice University in Texas, Kroto met with Robert Curl, an expert on spectroscopy, who showed him a highly powerful laser that could vaporize chemicals into a plasma of atoms. The device was created by a fellow Rice scientist Richard Smalley.

TIMELINE

1985	1991	1993
Curl, Kroto and Smalley discover a new carbon allotrope, the buckyball	Sumio Iijima elucidates the structure of carbon nanotubes	Fullerenes are found to inhibit the HIV-1 protease molecule

HAVING A BLAST

In a series of experiments, Kroto, Curl and Smalley used the laser to vaporize a sample of graphite. The carbon structures that formed in the vapour were analysed, and the team noticed a curious phenomenon. A previously unknown carbon allotrope, containing 60 atoms, formed in high abundance. Further research proved that in order to be stable, this allotrope must be spherical in shape, with 60 vertices, made up of 12 pentagons and 20 hexagons, just like a football. Each molecule was around 1 nanometre in size, which is one-ten-thousandth the diameter of a human hair.

The team named their find the 'buckminsterfullerene' after the American designer Richard Buckminster Fuller, who created geodesic domes with similar structures. The name was later truncated to 'buckyball'. Initially, scientists were sceptical about the discovery, because carbon had been so well researched it was hard to imagine that different allotropes could still exist. Further proof and the discovery of other spherical carbon molecules with 70, 76 and 84 atoms reinforced the team's findings. The group of allotropes became known as fullerenes and Kroto, Curl and Smalley won the 1996 Nobel Prize in Chemistry for their work.

> **THIS IS THE THIRD FORM OF PURE CARBON AND WE WERE COMPLETELY SURPRISED, AS WAS ALMOST EVERYONE ELSE, THAT THIS HOLLOW SPHEROIDAL CARBON CAGE MOLECULE EXISTED AT ALL.**
> Harold Kroto

BUCKYBALLS IN SPACE

Buckyballs exist on Earth in various forms. They have been detected in gas from burning candles and in rocks such as shungite and fulgurite. Proof of the existence of gaseous buckyballs in space came from NASA's Spitzer Space Telescope in 2010, and two years later astronomers detected solid grains consisting of stacked buckyballs around the binary star XX Ophiuchi. Some astronomers have even speculated that buckyballs from outer space could have brought carbon to Earth, by crashing into the planet on meteorites. Buckyballs may also explain one of the most baffling mysteries of astronomy.

2012	2013	2014
NASA's Spitzer telescope detects solid rather than gaseous buckyballs in space	Scientists discover how fullerenes can be used in drug delivery	Researchers reveal a cage-shaped boron fullerene consisting of 40 atoms

Stairway to heaven

Scientists have dreamed of building a space elevator for more than a century: having the ability to winch cargo up into space would be far cheaper and less damaging to the environment than using rockets. But finding a material strong enough to make a 35,000-kilometre (21,700-mile) cable that can support its own weight was believed to be impossible – even steel will break after just 30 kilometres (19 miles).

The development of carbon nanotubes breathed new life into the idea, as engineers realized they can make lighter, stronger cables that can hold firm over the vast distance required. The International Academy of Astronautics estimates that a space elevator could transport the same amount of cargo as the space shuttle every few days.

For a century, astronomers have observed gaps in the spectrum of light reaching Earth from other stars in the Milky Way. Known as diffuse interstellar bands (DIBs), these are believed to be caused by dust and other molecules absorbing light, resulting in an observable dimming of particular colours. In 2015, a team at the University of Basel announced their analysis of the light absorbed by buckyballs under space-like conditions. They found that the fullerenes absorbed light in line with the pattern of absorption seen in the DIBs. The study also suggested that buckyballs remain intact for millions of years and can travel huge distances in space.

CARBON NANOTUBES

Research into buckyballs led, in 1991, to the discovery of another related allotrope. Carbon nanotubes are the cylindrical cousins of buckyballs and resemble sheets of rolled up wire mesh. These molecules are minuscule (just a few billionths of a metre in diameter) but the bulk matter they create is 20 times as strong as steel and only half the weight of aluminium. Carbon nanotubes possess astonishing chemical and mechanical properties, which is why they are considered a great technology of the future. They can conduct heat faster than diamond (previously recognized as the best thermal conductor), and they conduct electricity four times more effectively than copper.

Fullerenes take various shapes: a carbon nanotube (left) is just a billionth of a metre in diameter and has excellent tensile strength; buckyballs (right) are spherical and have the same diameter as a nanotube.

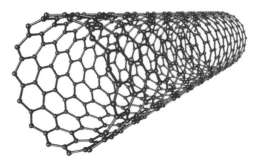

Nanotubes can be used as electrodes in capacitors because their large surface area enables them to store more charge than conventional designs. They are also being tipped to replace silicon in microchips to provide faster, more efficient access to data, via your personal computer, smartphone or smart watch. Thanks to their strength, structure and hard-wearing nature, they also make ideal building materials. Applications include sports equipment, such as golf clubs, body armour, rockets and construction materials. They are also being used to create mesh that filters out impurities, such as toxic chemicals and biological contaminants, to produce clean drinking water.

In medicine, nanotubes offer great potential in the treatment of cancer, where they can be used to deliver drugs to specific diseased cells in lower dosages. Fullerenes are known to be powerful antioxidants, slowing the production of harmful free radicals. Pharmaceutical companies are investigating their use to control neurological damage in Alzheimer's disease and motor neurone disease, and drugs to combat atherosclerosis and viral infections are also being tested. In 1993, American researchers found that fullerenes block the function of the HIV-1 protease, a key biological molecule in the progression of an HIV infection. It is now widely accepted that the field of HIV/AIDS research will be an important beneficiary of nanotube technology.

The discovery of fullerenes was one of the major contributors in opening up a whole new chapter of science – nanotechnology (see page 92). They are the engineering materials of the future and one of the most exciting developments of recent times.

The condensed idea
Engineering materials of the future

23 Nanotechnology

Scientists now have the ability to control and manipulate atoms to perform a vast range of functions. The invention of the scanning tunnelling microscope opened up a window on this nanoscale world which now offers the promise of exciting developments in medicine that could end the need for debilitating treatments, such as chemotherapy.

It's hard to fathom how tiny the realm known as the nanoscale really is. One nanometre (nm) is a billionth of a metre – equivalently, a millionth of a millimetre. For example, if a marble was 1 nm wide, in comparison the Earth would be 1 metre (40 in) across. Nanotechnology is the umbrella term that defines where biology, chemistry, physics and engineering meet on the nanoscale.

Nanotechnology is not new in itself. Nature is a master of the art and most biological processes take place on this scale. For example, haemoglobin, the molecule that transports oxygen in the body, is 5.5 nm in diameter and a DNA molecule is 2 nm wide. The science of nanotechnology operates on the quantum scale where materials behave differently from normal bulk matter. This opens up a whole new world of scientific and technological applications. Matter can become stronger, more chemically reactive, gain different magnetic properties and conduct heat and electricity more efficiently.

KEY DISCOVERIES

A number of breakthroughs led to the development of nanotechnology as a scientific field. The term was first coined in 1974 by Professor Norio Taniguchi

TIMELINE

1959	1974	1981
Feynman delivers a seminal talk highlighting the potential of controlling atoms	Taniguchi coins the term 'nanotechnology' to describe nanoscale engineering of materials	The scanning tunnelling microscope is invented, allowing scientists to view atoms

of Tokyo Science University to describe the ultra-precise machining of materials. This was followed by the invention of the scanning tunnelling microscope (STM) in 1981. Created by Gerd Binnig and Heinrich Rohrer at IBM Zurich, it enabled scientists to view surface features measuring just 0.01 nm across. Amazingly, this was precise enough to reveal individual atoms and their locations: a window had opened into the subatomic world and the duo won the Nobel Prize in Physics for their work.

From 1985, the discovery of nanoscale buckyballs (see page 88) led to the development of the carbon nanotube. This has proved key to the development of nanoscale electronics and devices, as well as new medical treatments. By the early 1990s, companies began launching products constructed out of nanoparticles onto the consumer market, including clear sunscreen, sports equipment, cosmetics, scratch-resistant glass coatings and improved displays for televisions and mobile phones.

Nanotechnology operates on a minute scale, yet it is immensely powerful. Break down matter into its tiniest form and you create a vast surface area, which boosts the reactivity of chemicals. Nanostructured materials make incredibly efficient catalysts. During the 2000s, some car manufacturers

Quantum tunnelling

Nanotechnology would not have developed without the invention of the scanning tunnelling microscope (STM) in 1981. The original electron microscopes, developed in the 1930s, allowed us to view particles or organisms down to one-thousandth of a millimetre across. The STM allows the study of objects 100,000 times smaller, on the nanoscale. It does this using a phenomenon of quantum physics known as 'tunnelling'.

In quantum physics, particles act like waves, and can pass through matter that classically they wouldn't be able to. With the STM, the tip of the probe is passed across the surface of a sample. It doesn't touch the sample but when voltage is applied, a current of electrons tunnels between the tip and the surface. If the surface gets closer to the tip or moves away, changes in the electric current are sensed and logged on a computer. This allows a super-detailed map of the atoms on the surface of the material to be created.

1985
Curl, Kroto and Smalley discover buckminsterfullerene, a carbon nanoparticle

1989
IBM scientists write their company's logo in xenon atoms using an STM

2015
The 7-nm microchip is announced – the smallest so far

Nanorobotics

The idea of millions of robots constructed on the nanoscale whizzing around the environment inspires either excitement or trepidation in people. Nanobots are devices that are programmed and controlled to work as a microscopic task force, destroying, repairing or constructing cells. For instance, if you catch a cold, nanobots could detect the offending virus and break down its atoms before you've developed so much as a sniffle.

The innovation has led to warnings of a doomsday scenario, where nanobots are released into the environment and start converting everything they come into contact with into copies of themselves. Within days, runaway replication occurs and the planet is turned into 'grey goo'. However, most scientists are sceptical this would ever happen.

began creating a new generation of catalytic converters that used platinum, rhodium or palladium particles just 5 nm wide embedded into ceramic plates. As a result, around 50 per cent less precious metal was required to drive the chemical reactions that convert pollutants into non-toxic emissions.

For many years, engineers have been tackling the challenge of how to cram more transistors onto microchips, to imbue personal computers and smartphones with greater speed and efficiency. The smallest features on most chips today are between 14 and 22 nm in size; however, in 2015, IBM announced that it had created a chip where the spacing is just 7 nm.

WORKING ON THE NANOSCALE

How do you control and manipulate atoms at this unfathomable scale? The invention of the STM allowed us not only to see tiny objects, but also to move them. In 1989, scientists at IBM famously used the technology to write the company logo in xenon atoms on a copper background. STMs with copper-tipped iridium wires can detect atoms and draw them along a surface into new positions.

The production of nanoscale materials is called nanomanufacturing. There are two approaches. Working top-down involves reducing bulk materials to the nanoscale, but this can be costly and produces a large amount of waste. Working bottom-up involves manufacturing materials from atomic and molecular components – a process that is incredibly time-consuming. As a result, scientists are exploring the possibilities of 'self-assembly', placing nanoscale molecules together so that they will grow from the bottom-up into ordered constructs but do so with the least possible outside direction. This process happens in nature all the time. For instance, water molecules self-assemble into ice crystals and fall to the ground as snowflakes.

MEDICAL APPLICATIONS

Nano-medicine has become a booming business in recent years. Scientists are now investigating the use of nanoscale gold in the treatment of cancer. Particles can be engineered to home in on tumours and gather within them, where they can be detected by precise imaging and then destroyed by lasers. Nanotechnology could also transform life for patients with diabetes. Researchers are developing a non-invasive device using nanotechnology that works like a breathalyser to detect acetone in breath, which correlates with blood sugar levels. Instead of having to use an uncomfortable finger prick several times a day, patients would only have to breathe into the device.

Nanotechnology has courted controversy over the years, with doubters warning that nanoparticles that we don't fully understand could enter the bloodstream and produce toxic effects. As with any other 'new' technology, research and testing are required to prove its potential – and to fully assess any dangers.

NANOTECHNOLOGY IN MEDICINE IS GOING TO HAVE A MAJOR IMPACT ON THE SURVIVAL OF THE HUMAN RACE.
Bernard Marcus

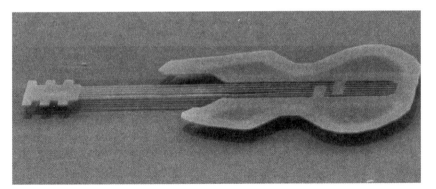

The world's smallest guitar, created by nanotechnology, is 10 microns long (about the size of a human red blood cell).

The condensed idea
Small but mighty science

24 The origin of life

One of the greatest mysteries of all is how life began on Earth. How did organisms develop following the geological chaos of our planet's early history? Did living matter originate in the oceans or did it hitch a lift from space on meteoroids and comets? The theories are as numerous as the questions.

The young planet Earth was a fierce environment. Debris raining down from space, meteorite impacts, volcanic eruptions and the decay of radioactive matter generated an intense cauldron of heat. It's no wonder that this epoch of the planet's history is called the Hadean era, after the Greek word *Hades*, meaning hell. No life on Earth's surface or the ocean floor could have developed until the end of this period, between 3.7 billion and 4 billion years ago.

Conditions on the planet then began to moderate as the surface cooled, creating a solid crust with a rocky terrain. Unicellular microbes, such as bacteria, were the first life forms to appear and we have evidence of this in the earliest fossils called stromatolites. How did these living organisms develop from the detritus of non-organic matter left in the post-Hadean era? That is the question that has intrigued scientists for centuries and led to the field of abiogenesis – how life sprung from basic chemicals. We still don't have an answer, only theories.

COOKING PRIMORDIAL SOUP
In the early 20th century, the scientists Alexander Oparin and John Haldane independently proposed the theory that in primordial times, before

TIMELINE

4TH CENTURY BCE	1903	1920s
Anaxagoras proposes that all living matter is derived from universal 'seeds'	Swedish scientist Svante Arrhenius speculates that spores can survive in space	Scientists propose life was created from the primordial soup activated by energy

Earth's atmosphere became oxygenated, a wide range of organic compounds could be generated if there was a supply of energy from sources such as lightning or ultraviolet radiation. In 1952, Stanley Miller and Harold Urey tested this theory in a famous experiment carried out at the University of Chicago. They simulated prebiotic conditions on Earth by combining water, hydrogen, methane and ammonia (the chemicals believed to have existed at the time) and pulsed this concoction with an electric current. A week later, the mixture had generated a number of organic compounds, including amino acids, the building blocks of proteins. Miller and Urey's findings supported the theory that the chemicals needed to form life could be created naturally on Earth.

Primordial soup

The notion of the primordial soup, and variations on the term, has been bandied around for almost a hundred years, but what does it actually mean? In 1924, the Soviet biologist Alexander Oparin proposed that Earth's early atmosphere was a heady mix of carbon dioxide, methane, ammonia, hydrogen and water. In this chemically reducing atmosphere (that is, with very little oxygen present) electrical activity could catalyse the creation of organic compounds, including amino acids – the building blocks of proteins. These compounds began to create other more complex molecules and eventually living organisms. British scientist John Haldane shared Oparin's view and coined the term 'prebiotic soup' for his vision of the sea as a hot, dilute chemical laboratory, where early life formed. Over time, the word 'prebiotic' evolved into 'primordial'.

Since the 1950s, further experiments have revealed that these prebiotic mixtures can also create nucleotides, the compounds that make up RNA (ribonucleic acid) and DNA (deoxyribonucleic acid) – the molecules that store our genetic code. This leads to a novel concept of how life may have started on Earth. The 'RNA world' theory was proposed in 1968 by Carl Woese who speculated that the early planet was home to abundant RNA produced in prebiotic chemical reactions. RNA doesn't just carry genetic code – it's also a catalyst, accelerating the rate of reactions without being changed itself. This would have been highly

1952

Scientists Miller and Urey recreate primordial soup and produce organic compounds

1968

American Carl Woese suggests that our earliest life forms were based on RNA

2014

International Space Station studies prove that bacteria can survive in space advantageous for producing multiple life forms in early times. But it's far from the only theory for how life got started.

> **THE FACT THAT LIFE EXISTS ON THE EARTH DOES NOT MEAN THAT LIFE STARTED ON THE EARTH.**
> Chandra Wickramasinghe

DEEP-SEA INCUBATORS?

In 1979, hydrothermal vents were discovered on the ocean floor near the Galapagos Islands in the Pacific. This led to the discovery of ecosystems containing fish, crustaceans, bacteria and organisms, which thrived far from sunlight in conditions previously believed to be uninhabitable. The heat and nutrients belched out by these 'black smokers', as they're also known, were capable of sustaining a hotbed of life. Scientists soon began to speculate whether life could have originated around these deep sea vents, with gaps in iron sulfide rocks becoming 'incubators' for primitive life forms. The jury is still out on whether life did originate on the ocean floor, but hydrothermal vents have been shown to create the perfect environment for it to happen.

A HITCHHIKER'S GUIDE

What if life on Earth didn't start on our planet at all, but in outer space? That would make us all aliens. The concept may sound like the stuff of science fiction, but perhaps it isn't as wacky as it seems. As early as the fourth century BCE, Greek philosopher Anaxagoras was the first to propose the concept of 'panspermia', suggesting that the universe originally consisted of an infinite number of seeds and that all life is derived from them.

Scientists in the 18th and 19th centuries began to speculate whether these seeds fell from space to Earth or were transported by meteoroids, asteroids and comets. Step forward the Swedish scientist Svante Arrhenius, who speculated in the early 20th century that life in the form of spores could survive space and be propelled through it by light pressure from the stars. This theory was supported in the later 20th century by the eminent astronomers Sir Fred Hoyle and Chandra Wickramasinghe, who believed that microbes enter our atmosphere all the time, causing outbreaks of disease.

Is there any proof that microbes could survive the rigours of space? In 2014, researchers on the International Space Station discovered that spore-forming bacteria that withstand extreme conditions on Earth are also able to survive a journey through space. These hardy bacteria have the potential to colonize other planets by hitching a lift on spacecraft.

NATURAL SELECTION

Some scientists refute the extraterrestrial and RNA theories, stating that Earth's primordial soup continued to react with itself, producing ever more complex compounds that, over time, generated life. Researchers at Massachusetts Institute of Technology (MIT) have taken this concept further and devised a model based on mathematical formulae and established physics. Their findings suggest that when atoms are surrounded by a 'heat-bath', in the ocean or in air, and driven by a source of energy like the Sun, they will restructure in order to dissipate energy more efficiently. MIT's researchers propose that this process of remodelling will inevitably lead to new life.

The concept touches on Charles Darwin's theory of natural selection – that organisms evolve in order to survive and reproduce more effectively in their environment. Further research and testing are required to prove whether MIT has made a leap forward in our understanding of abiogenesis. But for now the question of where did humans, and all other life on Earth, ultimately come from poses one of the greatest mysteries facing science.

The deep hot biosphere

In the 1990s, Austrian-born scientist Thomas Gold challenged the thinking about abiogenesis when he published his controversial 'deep hot biosphere' theory. Gold proposed that below the surface of the planet exists a biosphere of greater mass and volume than the sum total of living things on the surface. He believed that its inhabitants are heat-loving bacteria that feed on hydrocarbons such as methane gas (it has been proved that micro-organisms can survive underground to a depth of 5 kilometres or 3 miles). Gold went further – too far, some would say – to suggest that these bacteria were actually generating fossil fuels and that there is a huge reserve of oil being produced underground.

The condensed idea
How do you make life from base chemicals?

25 Photosynthesis

Plants are said to be the 'lungs of our planet' because they produce the oxygen that we breathe through a process known as photosynthesis. It is probably the most important chemical reaction on Earth and is part of a unique symbiosis between plants and living creatures.

Take a deep breath. The life-giving oxygen that you have just inhaled has been provided by green plants. Now breathe out. The carbon dioxide that you have exhaled is just as important for those plants as oxygen is for us. Photosynthesis is the complex process that plants use to create energy. Carbon dioxide and water are converted into sugars and oxygen is released as a by-product. It is the chemical reaction that ensures the survival of every living being on Earth – from the plants themselves, to herbivores that feed on them and predators searching for a well-fed catch.

We owe more than our existence in the here and now to photosynthesis. It enabled the very development of animal and human life on Earth. Without it, we would still be a planet of primordial soup.

The story of photosynthesis dates back billions of years to a major environmental change that scientists have called the oxygenation event. Prior to that, most of Earth's atmosphere was a smog of carbon dioxide produced by volcanoes. Around 3 billion years ago cyanobacteria (or blue-green algae) began producing oxygen through photosynthesis; however, most of this was used up in iron oxidation – literally rusting, as the oxygen combined with iron in rocks. Around 500 million years ago, land plants began to grow, generating even more oxygen, and the gas accumulated in the atmosphere at the level of around 21 per cent, where it remains today. This increase in atmospheric oxygen enabled the development of multicellular organisms such as humans and animals, which need oxygen to thrive.

I THOUGHT I WAS PRETTY COOL UNTIL I REALIZED THAT PLANTS CAN EAT THE SUN AND POOP OUT AIR.
Jim Bugg,
musician and writer

SOLAR POWER

The process of photosynthesis takes place in cells called chloroplasts. Plants draw up water through their roots and absorb carbon dioxide through tiny holes in the underside of their leaves called stomata. Light from the Sun is captured by the green pigment called chlorophyll, which uses the energy to drive two chemical reactions. On exposure to sunlight, the plant divides water molecules into hydrogen, oxygen and electrons. Further reactions combine these products with carbon dioxide to provide oxygen and carbohydrate. The plant feeds on the sugars of the carbohydrate and oxygen is released back into the atmosphere via the stomata. The equation for photosynthesis is:

$$CO_2 + H_2O + \text{sunlight} = CH_2O + O_2$$
(where CH_2O is carbohydrate, which the plant uses for energy)

Temperature is key in photosynthesis. If it is too hot or too cold, the plant will stop photosynthesizing. Increasing light intensity will accelerate the process, which is why some growers use either artificial light to extend photosynthesis beyond daylight hours, or a higher than normal light intensity to enhance the plant's production of its natural food.

The discovery of photosynthesis began in the late 18th century when the English chemist Joseph Priestley carried out an experiment proving that plants

1804
Nicolas de Saussure defines the role of water in photosynthesis

1881
Theodor Engelmann discovers that light energy is converted in the chloroplast

1956
Melvin Calvin discovers the chemical reactions plants use to synthesize sugars

Jan Ingenhousz (1730–99)

The Dutch biologist Jan Ingenhousz was also a doctor, having studied medicine at the University of Leuven. He was particularly interested in inoculation against smallpox and travelled to England in 1767 where he successfully inoculated 700 people in Hertfordshire to stop an epidemic. He was invited to the court of the Habsburg empress Maria Theresa, who had lost two relatives to the disease. Despite opposition from the Austrian medical establishment to inoculation, Ingenhousz injected small amounts of germs taken from a smallpox sufferer into his patients. The procedure was a success and he became court physician to Maria Theresa (mother of the ill-fated Marie Antoinette). In addition to the gaseous exchange of plants, Ingenhousz also investigated the processes of heat conduction and electricity.

produce oxygen. He placed a mint plant and a lit candle in a sealed glass container. The flame used up all of the oxygen and went out. However, Priestley was later able to relight the candle using magnified sunlight. This implied that the plant had produced more oxygen, enabling the candle to burn.

OXYGEN BUBBLES

Intrigued by Priestley's experiment, Dutch biologist Jan Ingenhousz placed plants in transparent containers underwater and noted that if sunlight fell on them, bubbles would appear on the underside of the leaves. If the plants were placed in darkness or shade, the bubbles eventually stopped. Ingenhousz found that the gas collected would make a candle flame burn far brighter, because the plant was producing oxygen. Ingenhousz is therefore credited with discovering the process of photosynthesis in 1779.

The next leap forward in our understanding of the process came in 1796 when Swiss naturalist Jean Senebier demonstrated that, under the influence of light, plants not only produce oxygen but also consume carbon dioxide. The key role played by water, meanwhile, emerged in the early 19th century when Swiss chemist Nicolas de Saussure observed that the carbon plants absorb from carbon dioxide could not account for the growth of the plant. He deduced that the increased weight was derived from water absorbed from the soil via the plant's roots. The mystery of the precise chemical process behind photosynthesis was starting to clear.

German botanist Theodor Engelmann discovered through a number of experiments in 1881 that solar energy is converted into chemical energy in the chloroplast and that only the red and blue components of sunlight activate these reactions. The term 'photosynthesis' was coined in 1893 by the American botanist Charles Barnes who also proposed the alternative 'photosyntax'. Barnes preferred the latter name, but it was photosynthesis that passed into common usage.

RADIOACTIVE TRACING

Developments in nuclear physics in the 20th century allowed researchers to nail down the pathways and processes of the key elements in photosynthesis. In 1941, American scientists Samuel Ruben and Martin Kamen traced the movement of oxygen through plants using a radioactive isotope. They proved that the oxygen produced in photosynthesis comes from water absorbed by the roots.

Rainforest destruction

Tropical rainforests are massive factories of photosynthesis. They cover only 6 per cent of the planet's surface, but at least 20 per cent of the oxygen on Earth is produced by their dense canopies, while huge amounts of carbon are stored in their vegetation. The rainforests are crucial in maintaining delicate ecological cycles and they influence global rainfall and climate patterns across the world. Burning, or cutting down rainforest vegetation and allowing it to rot, releases carbon dioxide into the atmosphere as CO_2 – whereas locking it away in living plants helps combat climate change. The current rate of rainforest destruction is around 32,000 hectares (80,000 acres) per day – pause for thought, if you consider that a standard football pitch is just over an acre in size.

In 1956, American chemist Melvin Calvin and his team made a huge breakthrough when they used a radioactive carbon isotope to follow the pathway of carbon in plants. This identified the chemical reactions plants use to convert carbon dioxide and other compounds into sugars. These became collectively known as the Calvin cycle. This final piece of detective work defined what is the most important chemical reaction on the planet, and Calvin was duly awarded the Nobel Prize in Chemistry in 1961.

The condensed idea
Green plants are the life force of our planet

26 The cell

Cells are the building blocks of life. The human body contains trillions of them, and yet we all originated from a single cell, the zygote, at the moment of conception. The cell is the smallest biological unit capable of replicating itself, a process that is rapid and ongoing in plants, humans and animals.

The word 'cell' is derived from the Latin *cella*, meaning a small room, and these tiny powerhouses of activity are the basic unit of all living organisms. They have a huge range of functions, from reproduction and growth to energy production and homeostasis (regulating conditions inside their parent organism, such as temperature, blood pressure, acidity and so on). Yet their size is tiny – it would take around 10,000 human cells to cover the head of a pin. Each cell can be viewed as an enclosed vessel, producing its own energy and replicating itself. Every human being loses 96 million cells per minute, but the same amount of cells divide simultaneously to replace them. Cells are in constant communication with surrounding cells and groups of similar cells combine to form tissue and vital organs.

CELL DISCOVERY
Early developments in microscopy led to the discovery of the cell in 1665. British scientist Robert Hooke observed slivers of cork through his microscope and noticed a honeycomb of irregularly shaped 'microscopical pores'. Hooke was actually looking at the plant's cell walls. The box-like structures reminded him of monastic cells and it was Hooke who coined the term. A century later, debates over the nature of cells began among

TIMELINE

1665	1670–80	1831
Robert Hooke discovers cells by studying cork through a primitive microscope	Van Leeuwenhoek uses a microscope to observe single-celled organisms	Botanist Robert Brown discovers the cell nucleus during a study of orchids

scientists. This led to 'cell theory', proposed in 1839 by the German biologists Matthias Jakob Schleiden and Theodor Schwann. The key characteristics were that cells are the basic unit of life, all living organisms consisted of single or multiple cells, and cells were capable of replacing themselves.

PROKARYOTES AND EUKARYOTES

There are two main types of cell. Eukaryotes are the most complex and make up the framework of multicellular organisms such as people and plants. Prokaryotes are the simpler form and they lack a nucleus – where the DNA resides in eukaryotic cells. Prokaryote DNA instead drifts freely in an internal space called the nucleoid, which lacks an outer membrane. All prokaryotic

> **THE BODY IS A COMMUNITY MADE UP OF ITS INNUMERABLE CELLS OR INHABITANTS.**
> Thomas Edison

organisms are single-celled and examples include bacteria and archaea, the earliest living organisms on Earth. If you've ever been unfortunate enough to suffer from salmonella or strep throat, your body has played host to prokaryotic organisms. Multicellular organisms are formed from eukaryotes. However, some unicellular organisms, such as amoebae, are also eukaryotes because they contain a nucleus bound by a membrane. These single-cell eukaryotic life forms are sometimes known as protists.

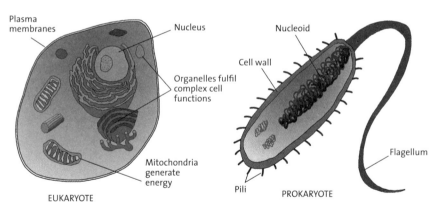

Eukaryotic cells (far left) have a nucleus, while prokaryotes (near left) carry DNA in a free-floating nucleoid.

1839
Cell theory is formulated by Matthias Jakob Schleiden and Theodor Schwann

1855
Rudolf Virchow proposes that all cells come from pre-existing cells

1857
Albert von Kölliker discovers energy-generating mitochondria in muscle tissue.

Reproductive cells

The cells that are involved in sexual reproduction are called gametes (the male are sperm and the female are ova). These differ from other cells in the body in a number of ways. In humans, the cells that form most of our body are diploid – containing two copies of each chromosome in the cell. Gametes are haploid, meaning they have only one copy of each chromosome. When a sperm cell fertilizes an ovum, the haploid gametes fuse to form a diploid zygote, the first cell of the offspring.

Gametes are formed through 'meiosis', a specialized form of division whereby the cell undergoes two fissions in order to produce four new cells instead of two. Meiosis occurs in eukaryotes and prokaryotes that reproduce sexually, and creates spores and pollen as well as animal gametes.

Eukaryotic cells contain a number of components, known as 'organelles'. Mitochondria control the cell's metabolic processes and generate energy from nutrients via the formation of adenosine triphosphate (ATP). Ribosomes are spherical structures that build proteins within the cell by linking chains of amino acids. The Golgi apparatus is often said to resemble a maze: its jobs include modifying proteins and lipids before delivering them to the required destination. Cavities known as lysosomes and vesicles, meanwhile, contain enzymes to drive specific chemical processes. All the organelles float around in the cytoplasm, the watery liquid that fills the cell, and are held in place by the cell's outer plasma membrane. Meanwhile, the cytoskeleton gives the cell its structure and shape.

The nucleus is the control centre of the eukaryotic cell and contains chromosomes where genetic code is stored on DNA. All the information the cell needs to perform its function is held on the DNA, whether it is a white blood cell fighting infection or a plant cell absorbing light via the chloroplast (see page 100). The nucleus is surrounded by what is known as the 'nuclear envelope', a porous membrane that allows free passage of proteins and RNA, which act as chemical messengers.

LIFE EXPECTANCY

Cells have a varying lifespan. For instance, those on the acid-rich surface of our stomachs are recycled every 5 days, while liver cells survive between 300 and 500 days and the human skeleton is renewed every 10 years. Cells are killed by bacterial infections, oxygen deprivation and poisoning. They can also be programmed to die off in a process called apoptosis, which is actually favourable to health and development. For example, the webbing between developing fingers and toes in the human embryo dies off by apoptosis to allow the digits to develop.

Cell division is a vital process for growth, maintenance and repair in a living organism. How a cell divides depends on its type. Prokaryotes undergo 'binary fission', where the cell divides in two, creating new 'daughter' cells. The tightly coiled DNA in the centre of the cell unfurls and replicates to form two copies. The DNA strands then pull to opposite poles of the cell, stretching the membrane until it breaks apart to form two new prokaryotic organisms that are identical to the parent cell.

The process by which eukaryotic cells divide is called mitosis. The double helix DNA strand 'unzips' along its length and the nucleotide bases on each half join with other bases to create two identical copies of the original strand. A process called cytokinesis then occurs, where the cell's nucleus divides in half, forming two new nuclei each containing one of the new strands of DNA. The rest of the cell then divides to create two new eukaryotic cells The whole process takes just over an hour from start to finish. Cell division in humans occurs continuously; however, it increases by 300 per cent during the early hours, which is why it's so important to get a good night's sleep.

Chromosomes

These long, thread-like structures in the nucleus of eukaryotes contain the DNA that passes genetic code from parents to offspring. Each coil of DNA is wrapped tightly around proteins called histones. Without these, the DNA would be too long to fit inside the nucleus. If all the molecules of DNA in just a single human cell were unwound and laid end to end they would stretch over 1.8 metres (6 ft).

Humans have 22 specific chromosomes and each cell in our body contains two copies of these, plus a pair of sex chromosomes that determine whether we are male or female. Our genetic make-up is contained on these pairs, including any abnormalities. For instance, red/green colour blindness is passed from mother to son on chromosome 23 (the sex chromosome) and early-onset Alzheimer's disease is caused by genetic mutations on chromosomes 1, 14 and 21.

The condensed idea
The building blocks of every living organism

27 Germ theory of disease

Millions of lives have been saved over the last 150 years through Louis Pasteur's discovery of germ theory. He proved that airborne micro-organisms cause disease and this revelation led to life-saving developments in medical practice, use of antiseptics and the pasteurization of food to kill dangerous bacteria.

The germ theory of disease began to emerge among physicians in the 16th century. Prior to that, diseases such as cholera and the bubonic plague were believed to be caused by 'miasma' – poisonous air containing fragments of decomposed matter that would infect anyone nearby. Early physicians also believed that disease was caused by spontaneous generation, whereby organisms such as fleas or maggots could form from inanimate matter like dust or decaying flesh. In 1668, Italian physician Francesco Redi put this theory to the test. He placed meat in three identical jars and left one open, another tightly sealed and covered the final one with gauze. Over time, Redi observed that the meat in the open jar became covered in maggots; the jar covered with gauze only had maggots on the surface of the gauze, and the tightly sealed jar had no maggots in it at all. Redi had sowed early seeds of doubt over the theory of spontaneous generation.

The true cause of disease became clearer following the development of the microscope in the 17th century. Dutch scientist Antonie van Leeuwenhoek

TIMELINE

1546	1668	1676
Physician Girolamo Fracastoro proposes that epidemics are caused by airborne spores	Francesco Redi disproves the 'spontaneous generation' theory of disease	Van Leeuwenhoek observes living cells through a microscope

became the first person to observe unicellular and multicellular micro-organisms, which he found in pond water. Later, scientists would speculate that diseases could be caused by worms or poisonous insects only visible through a microscope.

> **GENTLEMEN, IT IS THE MICROBES WHO WILL HAVE THE LAST WORD.**
> Louis Pasteur

The key breakthroughs in germ theory came in the 19th century, aided by insight into the nature of cells and how they divide. If disease was caused by cellular organisms, the process of cell division might explain how these organisms grow, thrive and colonise to cause epidemics. An outbreak of cholera on the filthy and overcrowded streets of 1850s London offered intriguing clues. Following a localized outbreak of the disease in Soho, physician John Snow identified the source of the disease as a public water pump near a cess pit on Broad Street. He persuaded the council to remove the handle of the pump and incidences of the disease declined. We now know that cholera is caused by the contamination of drinking water by effluent.

PASTEUR'S BREAKTHROUGH

It was the great French chemist Louis Pasteur who proved that micro-organisms cause disease, overthrowing any archaic notions of spontaneous generation. Pasteur did this by studying fermentation in nutrient broths, in a similar experiment to Redi's. He placed meat, sugar and water in two different flasks, one with a straight neck and the other with a swan-shaped neck. He boiled them to sterilize the contents and left them exposed to air. The broth in the straight-necked flask became cloudy with fermentation and was found to be teeming with microbial life. The broth in the swan-necked flask remained clear.

Pasteur concluded that airborne germs could fall directly into the straight flask, contaminating its contents. Germs entering the other flask had been caught in its curved neck, which is why it remained clear. Had spontaneous generation occurred, the broth in both flasks would have fermented. He

1830s	1864	1867
Cell theory enhances understanding of how organisms survive and spread	Pasteur makes history by proving that micro-organisms cause infection	Joseph Lister advocates the use of antiseptics in medicine to fight bacteria

The saviour of the mothers

Ignaz Semmelweis (1818–65) was a German-Hungarian physician who worked in the maternity clinic of Vienna General Hospital. In 1847, the unit had two clinics: one run by doctors and the other by midwives. Semmelweis was struck by the fact that mothers giving birth in the first unit were far more likely to die of fever than those delivered by midwives.

Semmelweis realized that unlike the midwives, the doctors were carrying out autopsies and passing on 'cadaverous particles' to the women. He instructed doctors to wash their hands in a bleach solution before entering the maternity ward, and fever deaths fell from 18 per cent to 2 per cent in just a month. Sadly, Semmelweis's belief in cleanliness was ridiculed by his contemporaries. He was vindicated posthumously by the findings of Pasteur and Lister.

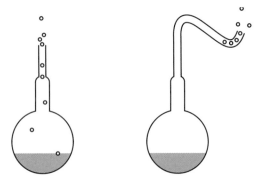

Pasteur's experiment with broth proved that airborne micro-organisms can infect food. Broth in the left-hand flask became contaminated, while that on the right remained clear because the curved neck of the flask trapped the germs in the air.

repeated his experiments using a filter to trap particles and again the broth remained clear. Airborne microbes must have travelled into the exposed jar to cause contamination. The conclusion was clear and historic: micro-organisms cause infection. They can be transferred through air, physical contact or via contaminated food and water. The germ theory of disease was accepted as a sound scientific principle in 1864.

Pasteur's discovery inspired other scientists and physicians who would revolutionize surgery and the treatment of disease. British doctor Joseph Lister read of the findings and applied them in medicine. Even in the 1860s, doctors were not required to wash their hands before entering the operating theatre or examining patients (see box, left). Many patients survived surgery only to fall victim to post-operative infections. From 1867, Lister began using carbolic acid as an antiseptic on wounds, surgical instruments and dressings. The incidence of infections, including gangrene, fell dramatically. Lister also instructed surgeons to wash their hands in a carbolic acid solution, a precursor of modern-day 'scrubbing up'.

MAGIC BULLETS

Now that the culprits of disease had been identified, scientists turned their attention to identifying specific germs and finding the chemicals that would destroy them. In the 1870s, Prussian doctor Robert Koch and his team devised a method of staining bacteria to make them easier to identify under a microscope. These pioneers discovered the bacterial causes of anthrax, cholera and consumption. Koch's methods also opened the door to other scientists who identified the bugs causing tetanus, typhus and plague. His colleague Paul Ehrlich, meanwhile, searched for a chemical that would not only stain a specific bacterial strain, but also kill it without causing harm elsewhere in the body. After hundreds of tests, Ehrlich and his team found their first magic bullet, the chemical arsphenamine. This destroyed the bacteria causing syphilis and was later marketed as the drug Salvarsan. Importantly, Ehrlich advocated the approach of treating the cause of disease rather than its symptoms.

The discovery of germ theory led to advances in food hygiene, including pasteurization, and the advent of antibiotic medication. Millions of lives were saved by the pioneering work of those who believed in germ theory while others remained sceptical. They take their place among the greats of scientific history.

Pasteurization

Louis Pasteur pioneered the heat treatment of milk and foods to kill off harmful bacteria. Raw milk can contain dangerous bacteria such as *E. coli*, salmonella, brucella and campylobacter. Before pasteurization, thousands of people died each year from drinking contaminated milk that had been transported long-distance from farms to cities.

Pasteur heated milk to between 60 and 100°C (140 and 212°F), which killed off most of the harmful organisms. Pasteur also studied wine contamination and discovered that it, too, was caused by bacteria. Pasteurization isn't commonly used in wine production now, because it affects the ageing process, but many of the items in our shopping basket can be safely stored for weeks, if not months, thanks to his revolutionary heat-treatment technique.

The condensed idea
Airborne micro-organisms cause many deadly diseases

28 Viruses

Viruses are tiny particles, far smaller than bacteria, that can cause deadly pandemics. They were discovered by Dutch microbiologist Martinus Beijerinck in 1898, and vaccination programmes have since helped to eradicate some of the most devastating viral diseases from the planet. Viruses are now being used in exciting new treatments against cancer.

We all know the misery of the common cold virus – runny nose, sore throat and bouts of deeply antisocial sneezing. Many people annoyingly complain that they are suffering from influenza (flu) when they simply have a cold: flu is also caused by a virus, but it is a far more deadly viral cousin. The 1918 outbreak of Spanish flu killed an estimated 50 million people worldwide, more than double the fatalities of the First World War. Even today, influenza can be a highly dangerous illness for the elderly and people whose immune systems are compromised. Viruses also cause Ebola, avian and swine flu, chickenpox and SARS (severe acute respiratory syndrome). While science has developed vaccines for a number of deadly viral diseases, such as polio and smallpox, the search for further treatments remains a priority.

VIRAL STRUCTURE
Viruses occupy a unique place in the natural world. They are minuscule in size, at least 50 times smaller than a bacterium and cannot be viewed through a standard microscope. It is impossible to class them as living organisms because they are unable to reproduce independently and exist in an inert state until they meet a host cell. They function, in essence, as parasites.

TIMELINE

1796	1864	1885
Jenner inoculates a child against smallpox, leading to vaccination programmes	Pasteur discovers airborne bacteria can cause life-threatening diseases	Pasteur successfully treats a rabies victim using a new vaccine

Transmission of a virus can occur in a number of ways. HIV (human immunodeficiency virus) is spread via sexual contact and exposure to infected blood. Influenza is propelled across workplaces, public transport and schools by coughs and sneezes. The gastrointestinal scourge of norovirus can be transmitted via contaminated food or contact with an infected person. Viruses that affect plants are often transported by insects.

> **THE VIRUS THAT CAUSES AIDS IS THE TRICKIEST PATHOGEN SCIENTISTS HAVE EVER CONFRONTED.**
> Seth Berkley

Structurally, viruses are very simple. They carry genetic code, either DNA or RNA, wrapped up in a protein coat called a capsid. Viruses spread insidiously by hijacking a host cell and using the cell's replication machinery to create new viral particles. When a virus meets a host cell it will pierce the cell wall, break down the outer coat and inject its own genetic code. The viral code then replicates inside the cell and is encased in capsids formed from amino acids. When this process is complete, the cell bursts open, dies and new viral particles are released to infect neighbouring cells.

DISCOVERY OF VIRUSES

Following Pasteur's great discovery of the germ theory of disease (see page 108) in the 1860s, scientists began to speculate about other forms of infectious agents. In the 1890s, Russian botanist Dmitry Ivanovsky studied a mystery disease that was causing great damage to tobacco plantations. He extracted juice from infected leaves and passed it through a Chamberland filter (invented by Charles Chamberland, a colleague of Pasteur). The filter had pores that were smaller than all known bacteria and yet after filtration the juice remained infectious.

In 1898, Dutch microbiologist Martinus Beijerinck examined Ivanovsky's experiment and deduced that the tobacco plants

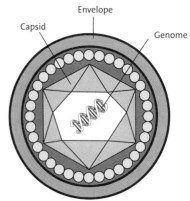

Viruses are simple particles comprising genetic material (the 'genome') wrapped up in a protein coat, or 'capsid'.

1892	1898	1931
Ivanovsky observes infection transferred by particles smaller than bacteria	Beijerinck discovers a 'new' infectious agent and calls it a virus	Invention of the electron microscope allows viruses to be seen for the first time

Discovering the smallpox vaccine

The eradication of smallpox from the planet is one of the biggest success stories in medical history. Caused by the *Variola* virus, smallpox is reputed to have killed more people over the centuries than all other infectious diseases combined. In the 18th century it was common knowledge that dairy workers didn't get smallpox (they were exposed to cowpox and developed antibodies that would fight both diseases). In 1796, the British doctor Edward Jenner tested the theory using his gardener's eight-year-old son James Phipps, who had never suffered from smallpox. Jenner inoculated the boy with matter from cowpox lesions. The boy became poorly for a few days but then recovered. Later Jenner inoculated him with smallpox matter. Crucially, James didn't go on to develop the disease. Vaccination programmes had eradicated smallpox worldwide by 1980.

were being attacked by a new kind of infectious agent. He called it a 'virus' – a name that had been in common use for centuries and was derived from the Latin word for poison. The disease affecting the plants was later found to be the tobacco mosaic virus. In 1931, the invention of the electron microscope by Ernst Ruska and Max Knoll enabled scientists to view viruses for the first time and witness their deadly processes. Many thousands of viruses were identified in the following decades.

IMMUNIZATION

The body protects itself from viruses by developing antibodies that recognize the protein patterns on the invader's capsid. They then destroy or consume the virus. Vaccines encourage the body to produce highly targeted antibodies. A dead or weakened form of the virus is administered and our immune system responds by producing white blood cells with a specific antibody that will destroy the pathogen. Children are usually immunized against viral illnesses such as polio, measles, mumps and rubella. The common cold remains impossible to immunize against because it is caused by viruses that mutate too rapidly. Some viruses mutate constantly, altering the patterns on their capsid, making it difficult for the antibody to recognize it.

Prevention is better than cure, but if a person is infected with a virus there is another line of attack. Antiviral medication can be used to treat a number of diseases, including herpes, hepatitis B and C, and also influenza A and B. They work by stopping the virus from replicating its genetic code, which limits the spread of the disease and buys time for the body's immune system to fight back.

VIRAL THERAPY AND CANCER

Viruses can cause some forms of cancer – however, a genetically modified virus is playing a pivotal role in a new cancer treatment. In 2015, a worldwide study found that a form of the herpes virus could be modified to destroy skin cancer cells. The virus (known as T-VEC) is introduced to diseased cells, replicates, then causes them to burst open. It also sparks the immune system to attack and destroy tumours. While healthy cells detect T-VEC and destroy it before it can do any damage, cancer cells cannot recognize it because they are already compromised.

Patients with stage three and early stage four melanoma who were treated with T-VEC lived for an average of 41 months, twice as long as the patients given current immunotherapy treatment. T-VEC has been hailed as a big leap forward in finding an alternative to radiotherapy, chemotherapy and surgery. After the carnage wreaked by diseases such as flu, AIDS and smallpox, it's refreshing to discover that not all viruses are our enemies.

The fight against Ebola

The 2014/15 outbreak of the Ebola virus in West Africa infected 27,500 people, killing at least 11,000 of them. It is believed that fruit bats of the Pteropodidae family are natural hosts of the virus, which enters the body through contact with blood, organs or the bodily secretions of carriers. Gorillas, porcupines and antelope can also carry Ebola. Consumption of these and other species as bushmeat may explain the transmission of the virus to humans.

The disease first appeared in 1976 in Zaire (now the Democratic Republic of the Congo), where it broke out in a village near the Ebola river. The virus spreads among humans by contact with bodily fluids and contaminated bedding and clothing. Fortunately, early trials of the vaccine VSV-EBOV in 2015 have shown promising results.

The condensed idea
Tiny parasitic agents cause infection and mutate to survive

29 Genes

Although DNA was discovered in 1869, the fact that it carries our genes wasn't understood until the mid-20th century. Genes hold the information needed to create every living organism on the planet. How they do this was first revealed by the work of a dedicated Austrian monk, experimenting with pea plants.

Genetic inheritance is a term for traits passed from parent to offspring via 'genes', units of information encoded into DNA molecules in the nuclei of cells. This happens across the natural world, from dogs to dahlias to human beings. Strands of DNA consist of sequences of genes, encoded on chemical bases. In the cells of an organism, these create a database of all the information needed to create a duplicate of that individual. Every cell in our body contains a copy of this database. Most genes work by encoding the information needed to make proteins, which are the major building blocks of complex organisms.

MENDELIAN TRAITS

We owe much of our understanding about genetic inheritance to an Austrian monk called Gregor Mendel. During his student days, Mendel studied the anatomy and physiology of plants under Franz Unger, a botanist and keen proponent of cell theory (see page 104). Mendel joined the Augustinian Abbey of St Thomas in Brno (in today's Czech Republic). In 1854, he began a meticulous seven-year study of the hybridization of the edible pea plant (*Pisum sativum*), noting traits such as plant height, flower colour and seed shape.

TIMELINE

1865	1869	1909
Mendel publishes his findings, that units pass traits from parent to offspring	Miescher discovers DNA through studying the nuclei of white blood cells	Johannsen coins the term 'gene' from the Greek word *genos*, meaning birth

First, Mendel ensured that he was using a 'pure' and identical line of plants, breeding them by self-pollination or cross-pollination for two years. It was an excellent baseline to work from. Two of the pea species grew into plants with either purple or white flowers. In one of the subsequent experiments, Mendel crossed the purple-flowering plants with the white-flowering variety. Instead of the flowers emerging with blended lilac tones, the first generation of plants all produced purple blooms. Mendel allowed this first generation to self-pollinate and intriguing results ensued. Some of the offspring had white flowers.

Mendel deduced that each plant carried two copies of the information for a trait, one from each parent, but that only one pure trait could actually be expressed – never a mixture of the two. This became known as the law of segregation. To explain why the flowers varied in colour, Mendel deduced that certain traits are 'dominant' (in this case, purple) while others are 'recessive' (in this case, white). Even if a plant carried both dominant and recessive factors, either could be passed to the offspring randomly (the law of inheritance). If both parents pass on the recessive factor then the resulting plant will bear that trait – which is how the white blooms reappeared. Mendel counted the number of plants with white and purple blooms through the generations. The ratio was 3:1, with purple being the dominant colour.

The ginger gene

If a redhead pops up in a family, where mum and dad have dark or blonde hair, it can seem incongruous. Red hair is unusual, because it is caused by a recessive form of the gene MC1R, found on chromosome 16. To have red hair, a child needs to receive a copy of this recessive gene from both parents (let's call the recessive gene r and the dominant 'non-red' gene R).

If neither parent has red hair but they both carry the recessive gene (Rr), the child has a 25 per cent chance of being a redhead. If, say, mum is a redhead (rr) and dad is not but is a carrier (Rr), the likelihood increases to 50 per cent. In families where both mum and dad are redheads (rr) all their children will follow suit, creating a colourful clan.

1911
Morgan discovers that traits can be carried on sex chromosomes

1928
Frederick Griffith observes that characteristics can be transmitted between cells

1944
Oswald Avery and his team prove that genes are carried on DNA

WE ARE SURVIVAL MACHINES – ROBOT VEHICLES BLINDLY PROGRAMMED TO PRESERVE THE SELFISH MOLECULES KNOWN AS GENES.

Richard Dawkins

Mendel called the inheritance traits 'factors', but they are what we now know as genes. Characteristics caused by a single gene, as in the case of the peas, are called Mendelian traits. Science has proved that genetic inheritance can be far more complex – however, Mendel had demonstrated two key factors: how characteristics are passed from parent to offspring and how natural variations occur within species.

DNA CARRIES GENES

Mendel published his findings, but like so many innovators he didn't gain recognition during his lifetime. Meanwhile, in 1869, the Swiss biologist Friedrich Miescher became the first person to isolate DNA, during a study of the nuclei of white blood cells found in pus. A series of experiments in the 20th century would make the crucial link between DNA and genes.

The term 'gene' was first used in 1909 by Dutch botanist Wilhelm Johannsen and is derived from the Greek word *genos*, meaning birth. However, the precise mechanism by which genes were passed from parent to offspring remained a mystery. In 1928, British microbiologist Frederick Griffith injected laboratory mice with two strains of pneumonia bacteria – one fatal, the other non-fatal. As expected, the mice injected with the fatal strain died and the mice injected with the non-fatal strain survived. Griffith then killed the fatal strain and introduced the dead bacteria to the mice, who survived. The experiment took an astonishing turn, however, when Griffith gave mice both the killed fatal strain and the living non-fatal strain of bacteria. The mice died and Griffith deduced that something had been passed from the dead fatal strain to the living non-fatal strain. How could this happen?

In 1944, a team of American scientists, Oswald Avery, Colin MacLeod and Maclyn McCarty, solved the mystery in a landmark experiment. Scientists had an inkling that proteins transmitted genes from one life form to another, and Avery's team tested this by using an enzyme to destroy the protein in the pneumonia strains. The mice still died, meaning the protein wasn't responsible. But when a DNA-digesting enzyme was introduced to the fatal strain, the mice survived. Avery thus knew that DNA was the molecule transmitting information between the two strains of bacterial cells. Later experiments by another team used radioactive tracer atoms to provide further proof that DNA carries genes. Avery was awarded the Copley Medal of the Royal Society of London in 1945 for his discovery, but was inexplicably passed over for a Nobel.

Forging the link between DNA and genes has led to the development of epigenetics, the study of how genetic code can be altered during our lifetime by factors such as environment and lifestyle. It has also led to gene therapy, an experimental technique that involves modifying a patient's genes to treat diseases such as cancer, Parkinson's disease, diabetes and AIDS.

Let's talk about sex

Pioneering American geneticist Thomas Hunt Morgan made history by showing that genes are linked in series on chromosomes and are responsible for definite hereditary traits. In 1907 he began studying the fruit fly (*Drosophila melanogaster*) in what became known as the 'fly room' at Columbia University. Among many experiments, Morgan found that male fruit flies could be born with white eyes (instead of the usual brilliant red). He crossed white-eyed males with red-eyed females and all the offspring had red eyes. The white eye reappeared in the next generation, but only in male fruit flies. Morgan had discovered that some traits were sex-linked and that the gene for them was probably carried on the sex chromosome. A new era in the understanding and study of genetics had begun. Morgan won the Nobel Prize in 1933.

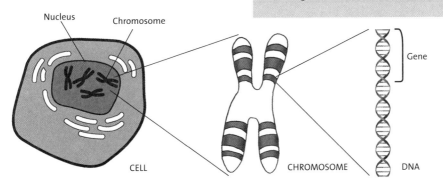

The nucleus of a cell contains chromosomes made up of DNA, which carries our genes.

The condensed idea
Genes carry the information code of life

30 Evolution

The cornerstones of evolutionary theory were laid by Charles Darwin after his legendary journey around South America's coast. At the time of Darwin's birth, most people believed that God created the Earth and everything on it. Evolution was his life's work and is among the most significant contributions ever made to science.

In the 18th century, naturalists began to question the idea that life had been fixed since creation. The study of palaeontology, pioneered by French naturalist Georges Cuvier, provided physical evidence of extinct species, and a picture of the natural world as an ever-changing environment. Erasmus Darwin, grandfather of Charles, proposed that life evolved from a common ancestor, but wrestled with the idea of how one species could give rise to another.

In the early 19th century, French botanist Jean-Baptiste Lamarck proposed a grand theory of evolution. He believed that if an organism used a characteristic to its advantage during its lifetime, this feature would grow bigger and stronger and be passed on to its offspring (so, for example, if you do a lot of physical work then you will have children with large muscles!). Some aspects of Lamarckism are now re-emerging through the study of epigenetics (how genes can be switched on and off by external influences). However, Lamarckism was insufficient to bring about the degree of change observed in the natural world, and was widely lampooned at the time.

Clarity would only come in 1859, when Charles Darwin published his controversial and seminal work *On the Origin of Species*. This defined his

TIMELINE

1794	1796	1801
Erasmus Darwin suggests all living organisms evolved from a common ancestor	Georges Cuvier, pioneer of palaeontology, publishes seminal theories on extinction	Lamarck proposes that environment influences characteristics and heredity

theory of evolution, the process by which organisms adapt to their environment in order to survive. It also explained how, over time, organisms could evolve so significantly that they give rise to entirely new species.

MY PERSONAL FEELING IS THAT UNDERSTANDING EVOLUTION LED ME TO ATHEISM.
Richard Dawkins

VOYAGE OF DISCOVERY

In 1831, Darwin was invited to join the crew of HMS *Beagle* on a survey mission around the coast of South America. Its captain, Robert FitzRoy, was a prescient man who realized it would be useful to have a naturalist on board. Darwin was aged 22 at the time and had become a keen amateur naturalist at Cambridge University. He leapt at the chance, and the *Beagle* embarked on its five-year voyage from Plymouth harbour on 27 December.

Darwin became an avid collector of fossils en route, fascinated by relics that seemed similar but different to modern-day species. For instance, in Argentina he found the skeleton of a horse-like creature with a long face similar to an anteater. The young naturalist began to ponder what caused these fantastic creatures to die out and to question the stability of species. Could they change over time? Darwin returned to Britain in 1836 with more than 5,000 specimens of birds, mammals, fossils and bones. He settled into life as a gentleman geologist and over the next 20 years formulated his landmark theories.

TAKE FOUR FINCHES

Darwin's study of finches was to unlock one of the key concepts of evolution. During his time on the Galapagos Islands, off the western coast of Ecuador, Darwin observed finches from each of the four islands. Back in Britain, he continued his study noting that all of the birds were fundamentally similar, apart from the shape of their beaks. The distance between the archipelago's islands meant that the finches could not cross-breed.

1859	**1865**	**1944**
Darwin makes history by publishing his theory of natural selection	Genetic inheritance is proved through Mendel's experiments on pea plants	DNA is proved to be the medium that passes on genetic information

Superbugs

The emergence of antibiotic-resistant bacteria, such as MRSA (Methicillin-resistant *Staphylococcus aureus*), is a darker example of natural selection. MRSA has become a problem through its stunning ability to mutate: one *S. aureus* bacterium can divide and create 300 mutations overnight, enabling it to rapidly adapt to threats such as antibiotics. In fact, overuse of antibiotics has driven the rise of these superbugs, by creating a selection pressure that gives them an advantage over less resistant strains. In addition to its chromosomal DNA, MRSA contains rogue elements of DNA called plasmids, whose genes make toxins that bind to antibiotics and inhibit their action. Plasmids can be exchanged between different strains of bacteria through 'horizontal gene transfer', spreading antibiotic resistance even further.

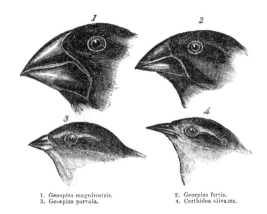

1. Geospiza magnirostris.
2. Geospiza fortis.
3. Geospiza parvula.
4. Certhidea olivasea.

Variation between the beaks of finches found on the Galapagos Islands helped Darwin formulate his theory of natural selection.

The finches fed off cacti, and Darwin realized that their beaks varied according to how they fed. Those with long beaks were able to punch a hole in the leaves to eat the fleshy pulp, whereas those with shorter beaks could tear at the tough base of the plant, ingesting insect life as well. As Darwin said, 'it was as though one species had been taken and modified for different ends'. The finches were later confirmed to be separate species in their own right.

NATURAL SELECTION

Darwin wondered how variation in the birds' beaks could occur and this led him to the theory of 'transmutation'. While Darwin disputed a number of Lamarck's ideas, he was intrigued by the idea of favourable traits passed from parents to offspring. He quizzed owners of fancy pigeons and dogs to discover how they enhanced slight variations in the creatures through breeding.

One of Darwin's key realizations was that organisms of a given species are in competition for scarce food and resources. Traits passed from parents to offspring boost their ability to compete, survive and reproduce, so favourable traits spread through a community and transform it over time. Darwin called this mechanism 'evolution by natural selection': we now know that the variety of traits on which it works arise from random mutations in DNA.

On the Origin of Species was published in 1859, and became a runaway success. Inevitably, it sparked controversy among the religious community – something Darwin anticipated when he commented that believing in natural selection was akin to 'confessing to murder'. Reaction within the scientific community was also mixed at first, but eventually the work of other naturalists and scientists, such as the rediscovered experiments of Gregor Mendel (see page 116), began to offer convincing evidence that parents pass genetic code to their offspring. By the middle of the 20th century, Darwin's findings were the benchmark for modern evolutionary study.

There are countless examples of natural selection all around us. Modern studies have shown that the finches of the Galapagos Islands react to weather changes, as well as food availability. After periods of drought, some species develop thicker, stronger beaks that allow them to tear through tougher seeds. Darwin's finches continue to prove the governance of his legendary theory, often known as 'survival of the fittest', long after his death.

Charles Darwin (1809–82)

Born into a wealthy Shropshire family, Darwin loved to explore nature from an early age. He was sent by his father to study medicine at Edinburgh University, but he hated the sight of blood and dropped out after his second year. His father then decided on a clerical career and Charles entered Christ's College, Cambridge, where he spent much of his free time studying nature, as well as riding, shooting and drinking.

Darwin became a more serious character following his five-year voyage on HMS *Beagle*. His intellectual circle included inventor Charles Babbage, geologist Charles Lyell and biologist Thomas Huxley. Darwin married his cousin Emma Wedgwood, daughter of the pottery magnate Josiah Wedgwood. They had ten children and settled at Down House, Kent. Darwin died in April 1882, and is buried in Westminster Abbey.

The condensed idea
Organisms adapt to their environment

31 Out of Africa

It's an astonishing thought that all human beings are descended from a common ancestor, who lived in Africa 200,000 years ago. The 'Out of Africa' theory of man's evolution was first proposed by Charles Darwin in 1871, and has been backed up by fossil finds and advances in our understanding of human DNA.

In his masterpiece *On the Origin of Species*, Charles Darwin gave a tantalizing glimpse of what was to come when he stated that 'light will be thrown on the origin of man and his history'. Darwin had already revolutionized conventional wisdom by proposing that species found in nature are not immutable, and that they evolve through a process of natural selection. Why shouldn't the same philosophy extend to mankind?

Prior to Darwin, some anthropologists believed that humans had multiple origins and that races emerged separately across the globe. In 1871, Darwin countered this in his book *The Descent of Man*, in which he proposed that all humanity had a common line of descent and that it originated in Africa.

A VISIT TO THE ZOO

Darwin lived in London following his journey on HMS *Beagle*. In 1838, he visited London Zoo and saw an orangutan, Jenny, who left a lasting impression on him. He commented: 'Let any man see an Ouranoutang... its intelligence... its affection... its passion and range, sulkiness and very actions of despair... then let him boast of his proud pre-eminence'. And he wasn't the only one to notice the similarities: Queen Victoria found the ape 'disagreeably human'.

TIMELINE

1800s	1871	1924
Anthropologists speculate that humans evolved from multiple origins	Darwin states man is descended from apes and evolved in Africa	The African Taung Skull links apes and early hominids

So it was that Darwin rocked the Victorian world yet again by proposing that humans had descended from apes – 'man's nearest allies' – and that this evolution occurred in Africa, which was home to gorillas and chimpanzees. The evolved species then drifted across the continents, adapting by further evolution to suit their living environment and creating new races in the process.

THE MISSING LINK

Few fossils of early mankind (hominids) had been discovered by the 1870s. But Darwin predicted, accurately, that relics of early humans and their progenitors would be found in Africa. He was proven right in 1924, when Raymond Dart, an Australian professor of anatomy based in Johannesburg, obtained a fossil skull unearthed near the town of Taung in South Africa.

The mystery of Piltdown Man

In 1912, newspaper headlines screamed the words 'Missing link found'. A fossil skull and jawbone were presented to the world by amateur British archaeologist Charles Dawson, who found them alongside animal fossils in a pit near Piltdown in Sussex. Piltdown Man was claimed to have lived 500,000 years ago and represent the link between apes and human beings.

Around 40 years later, scientists revealed that the skull was only 500 years old and the jawbone belonged to an orangutan. All of the other fossils at the pit were found to have been planted. Dawson died in 1916 and no one knows if he was the culprit behind this famous hoax, or whether a third party was to blame.

Dart knew he was on to something because the skull was too large and the teeth too small to be that of an ape. He also noticed that the foramen magnum (the hole through which the spinal cord passes to join the brain) was towards the front of the skull, an adaptation found in bipedal humans, which enables the head to balance on top of the neck. However, at 2–3 million years old, the 'Taung Skull' was much older than bones of Neanderthals and *Homo erectus* (other human species discovered in Europe and Asia in the 19th century). The relic was clearly intermediary between apes and human beings, and Dart dubbed his new find *Australopithecus africanus* ('southern ape from Africa').

1974
The 3-million-year-old skeleton of 'Lucy' is found in Africa

1987
Mitchondrial DNA links all mankind to 'Eve', a common African ancestor

2014
Y-chromosomal Adam is estimated to have lived 208,300 years ago

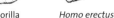

Gorilla   Homo erectus   Homo sapiens

Early hominids such as *Homo erectus* had larger brains and less protruding jaws than apes. *Homo sapiens'* skull was larger than *H. erectus* and lacked the prominent brow

Further finds, such as the skeletal remains of 'Lucy', a human ancestor found in Ethiopia in 1974, substantiate the 'Out of Africa' theory. Lucy lived around 3.2 million years ago and her pelvis, ankle and knee bones suggested that she had an upright, bipedal gait. Walking was clearly one of the most important selective traits that drove human evolution forwards, and studies of food remains found in the teeth of early hominids suggests that their diet was changing. As well as foraging for fruit in trees, they were turning their focus to grasses and possibly meat. Upright walking allowed them to scavenge further for food and protected their bodies from sun damage.

The Hobbit

In 2003, anthropologists discovered a curious set of skeletal remains on the island of Flores in Indonesia, the like of which had never seen before. The adults stood only 1 metre (40 in) tall and had long arms and a chimpanzee-sized skull, but they also had delicate bones, non-projecting faces and small teeth – traits normally seen in humans. On this basis, scientists declared it a new hominid species and called it *Homo floresiensis*. It also acquired the nickname 'The Hobbit'.

This unique community is estimated to have lived as recently as 19,000 years ago and may have been descended from the hominid species *H. erectus*. Speculation over their diminutive stature led scientists to consider 'island dwarfing', where creatures confined to isolated areas can become smaller over time due to limited food supplies.

MITOCHONDRIAL EVE

Scientists now believe that we all share a common ancestor, a woman who lived in Africa around 200,000 years ago. Evidence for this comes not from fossils, but from a source known as mitochondrial DNA.

The DNA in the nuclei of our cells, which determines our physical traits, mutates significantly as it is passed down through the generations. However, there is another form – human mitochondrial DNA (mtDNA) – which contains 37 genes that do not change. MtDNA is found within organelles called mitochondria (see page 116) responsible for synthesizing energy from food. We inherit it purely from our mothers because sperm mtDNA is destroyed after fertilization (it is 'matrilineal'). This allows us to trace our female ancestors back through time.

In 1987, a study of people from all major races found that their lineage fell onto the branches of two genealogical trees, one of purely African lineage, and the other containing lineage from other continents besides Africa. The team behind the study interpreted this to mean that all humans have a common ancestor from Africa – who they dubbed Mitochondrial Eve. Alongside the stable genes, mtDNA also contains a region that mutates. Measuring the rate of mutation, the team deduced that Eve must have lived some 200,000 years ago – meaning she co-existed with hominids such as Neanderthals and *Homo erectus*. This is not to infer that Mitochondrial Eve was the only female alive at the time. She is simply the most recent female ancestor to whom we can all trace an unbroken matrilineal line.

Similarly, the Y (male) chromosome is passed only from father to son (it's 'patrilineal') and is also largely unchanged from one generation to the next. Evidence from its analysis suggests that all men are descended from one common male ancestor 'Y-chromosomal Adam', who is now believed to have lived in Africa during the same era as Mitochondrial Eve.

The name of our species, *Homo sapiens*, means 'wise man'. Our ancestors lived alongside Neanderthals and *H. erectus* for thousands of years, but survived when other hominids died out because our larger brains adapted to provide the ability to plan, develop language and communicate ideas to others. This intelligence allowed us to craft better tools and weapons, which helped in the search for food – another triumph for the idea of natural selection.

DARWINIAN MAN, THOUGH WELL-BEHAVED, AT BEST IS ONLY A MONKEY SHAVED!
W.S. Gilbert

The condensed idea
All humankind evolved from African apes

32 The double helix

Cambridge scientists James Watson and Francis Crick made history by discovering the double helix form of DNA. Their revelation helped shine the light on how DNA replicates and transmits genetic code. This has led to breakthroughs such as gene therapy, offering hope of treatment for incurable diseases.

In 1953, Cambridge University scientists James Watson and Francis Crick solved one of the biggest riddles in natural science – the structure of DNA. Their three-dimensional model of the double helix transformed our understanding of the living world forever and laid the foundations for the study of genetics and molecular biology. Scientists had suspected for some time that DNA was the 'molecule of life', carrying the genes that make up our personal code. Oswald Avery and his team at the Rockefeller Institute, New York, proved this in 1944 (see page 116); however, the molecular structure of DNA – and the mechanics of how it stores our genetic code – still remained a mystery.

THE DNA SAGA

In 1869, Friedrich Miescher discovered DNA, which he isolated from the nuclei of white blood cells. Around the turn of the 20th century, the basic building blocks of the DNA molecule were known to be sugar, phosphate and four other chemicals – thymine, guanine, adenine and cytosine (usually labelled according to their initials T, G, A and C). These are known as nucleotides.

Avery's discovery of DNA as the transmitter of genetic information inspired Austrian biochemist Erwin Chargaff to analyse nucleotide bases in DNA

TIMELINE

1869	1944	1949
Miescher discovers DNA in the nuclei of white blood cells	Avery proves that DNA carries genetic code in organisms	Chargaff deduces the ratio of chemical components in DNA

samples. His chemical tests revealed that in any given sample of DNA, the proportion of A was equal to the proportion of T, while the proportions of G and C were equal, too. It seemed as though pairing was occurring between these nucleotide bases and these findings became known as Chargaff's rule.

THE CAMERA NEVER LIES

The scene now shifts to Cambridge University in the late 1940s, where British biophysicist Francis Crick was working in the Cavendish Laboratories on molecular structure. Here he was joined by American biologist James Watson, who had recently completed a post-doctoral study of virus DNA in Copenhagen. The duo were tasked to research the structure of DNA by the head of the laboratory, Sir Lawrence Bragg. An early attempt in 1952 missed the mark with a three-sided, inside-out construct that was quickly rejected. Hot on their heels, meanwhile, was American chemist Linus Pauling, who presented his triple-helix model in 1953. The pressure was on for Watson and Crick to claim the prize for Cambridge.

Rosalind Franklin, a chemist at King's College, London, provided one of the most important clues. She specialized in producing X-ray diffraction images, shining high-frequency X-rays through crystalline DNA, which diffracted through the molecules in the crystal lattice, to reveal their arrangement.

Franklin's pictures were becoming clearer and clearer, and in 1952, she produced an image of a DNA molecule

AT LUNCH FRANCIS WINGED INTO THE EAGLE TO TELL EVERYONE WITHIN HEARING DISTANCE THAT WE HAD FOUND THE SECRET OF LIFE.

James Watson
(The Eagle is a pub in Cambridge)

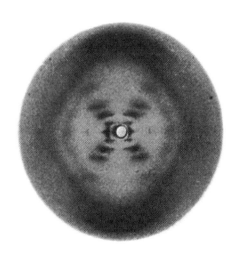

Rosalind Franklin's iconic 'Photo 51', taken in 1952, revealed DNA's double helix structure, and inspired Watson and Crick to build their 3D model.

1952	**1953**	**1962**
Rosalind Franklin's 'Photo 51' shows DNA's double helix shape	Watson and Crick build the first-ever 3D model of DNA	Wilkins, Watson and Crick win the Nobel Prize

Watson and Crick

Francis Crick was born in Northamptonshire in 1916. He began his academic career at University College, London, studying for a BSc in Physics. During the Second World War, he worked on the development of magnetic mines for use in naval warfare. In 1947, he turned his attention to biology, at Cambridge, focusing chiefly on the three-dimensional structure of molecules.

James Watson, born in Chicago in 1928, had a PhD in zoology. Watson and Crick hit it off at Cambridge, sharing a desire to fathom the molecular structure of DNA.

Following their landmark discovery, Watson became professor of biology at Harvard University, where he continued research into nucleic acids and protein synthesis. At Cambridge, Crick went on to explore the biological implications of DNA, including the process in which RNA (ribonucleic acid) copies DNA code and transports it into ribosomes where it creates proteins. Crick died in San Diego, California, in 2004.

called 'Photo 51'. The molecule showed a distinct X-shape, and suggested that DNA had a double helix structure, with two long strands wound round each other (the 'X' was formed by the two strands crossing over, as viewed from the side). The image also showed the nucleotide bases linked up across the two helices, like steps in a long spiral staircase. This substantiated Chargaff's rule regarding equal proportions between nucleotide bases in DNA: an A base on one helix linked to a T base positioned opposite, and likewise for C and G bases. A colleague of Franklin's, Maurice Wilkins, showed Photo 51 to James Watson. It was a seminal moment for the young scientist, who later said: 'the instant I saw the picture my mouth fell open and my pulse began to race.'

BUILDING THE DOUBLE HELIX

Yet despite having such a good template to work from, the exact chemical structure of DNA remained to be elucidated. Watson and Crick began building their model using metal plates for the nucleotide bases and rods for the bonds between them. Their 2-metre (80-in) structure was precise and complex, illustrating the angles formed by different chemical bonds. Crick's weighty mathematical prowess underpinned their calculations. The model was finished on 7 March 1953. Many great scientists had paved the way, but Watson and Crick had won the race to understand DNA's double helix.

The model also gave vital hints illustrating how DNA replicates. As they stated in their *Nature* paper of April 1953, 'It has not escaped our notice that the specific pairing we have postulated suggests a possible copying mechanism for the genetic material.' Crick later proved that DNA replicates by unwinding into two separate strands that each gain complementary nucleotide bases (A bonding to T, C to G, and vice versa) to form two new and identical double helices.

Watson, Crick and Wilkins were awarded the 1962 Nobel Prize in Medicine or Physiology. Tragically, Rosalind Franklin had died four years earlier of ovarian cancer and since the Nobel Prize is not awarded posthumously, she was not honoured. Crick made further significant breakthroughs, unravelling how DNA encodes genetic information and makes proteins. Watson wrote the bestselling book *The Double Helix*, which was acclaimed as a personal account, but criticized for sexist references to Franklin.

Gene therapy

Gene disorders account for around 4,000 diseases, including cancer, AIDS, Alzheimer's disease and cardiovascular disease. Gene therapy works by repairing dysfunctional genes or adding copies of those that are missing. It is most likely to be successful in treating diseases caused by single gene mutations, such as cystic fibrosis. The process works by isolating normal DNA and packaging it into a vector, often a virus. This is introduced to a cell affected by the disease, where the new DNA is unloaded and starts to do its job, creating the correct proteins needed for the cell.

Gene therapy can also target germ cells (eggs and sperm), with the aim of passing on disease resistance to the next generation. The latter is controversial and is banned in many countries.

The discovery of the double helix structure heralded a new world of scientific study – molecular biology. This has led to pioneering fields such as genetic engineering, new techniques in forensic science and gene therapy, which has the potential to treat some otherwise incurable illnesses. The legacy of all the protagonists in the DNA story continues to touch our lives.

The condensed idea
The structure of life's master molecule

33 Cloning and GM

Scientific advances in cloning and genetic modification have offered solutions to a number of pressing global concerns, including disease and malnutrition. In 1972, biologists worked out how to chop up DNA and recombine it, offering endless possibilities. However, the ethics and safety issues surrounding such experiments continue to cause division.

In 1996, scientists at Edinburgh University hit the headlines when they announced to the world the birth of 'Dolly', the first mammal to have been cloned from an adult body cell. Dolly sparked heated debate, with many people reflecting on the consequences of 'interfering with Mother Nature'. Cloning is essentially asexual reproduction, where identical offspring are produced by one parent. It has been going on for billions of years in the natural world, in bacteria, fungi and plants. Any keen gardener will know that plants can be cloned by taking cuttings that are potted on to create an identical new plant.

HELLO DOLLY

Scientists began experimenting with cloning around the turn of the 20th century. In 1928, German embryologist Hans Spemann separated the cells of a two-celled salamander embryo and successfully produced two larvae. In 1958, British biologist John Gurdon cloned a frog, using intestinal cells from an adult African clawed frog. However, Dolly the sheep was the first large mammal to be created through cloning, and, crucially, from an adult cell not an embryonic cell. She was the dream child of Ian Wilmut and his colleagues at the Roslin Institute of Edinburgh University.

TIMELINE

1928	1958	1972
Hans Spemann clones a salamander by splitting a two-celled embryo	John Gurdon clones an African clawed frog using intestinal cells	Berg is the first to make recombinant (modified) DNA from viruses

Wilmut used a technique called somatic cell nuclear transfer (SCNT) to create Dolly. An egg (germ) cell is taken from the mother animal and the nucleus of that cell containing all her genetic code is removed and discarded. A donor cell is then taken from the animal that is being cloned – a somatic (body) cell, not a reproductive cell. The nucleus of this cell is then implanted in the 'empty' egg cell where it begins to divide and becomes a blastocyst (containing around 100 cells). The blastocyst is then implanted into the mother animal where it continues to grow.

> WITH GENETIC ENGINEERING, WE WILL BE ABLE TO INCREASE THE COMPLEXITY OF OUR DNA, AND IMPROVE THE HUMAN RACE.
> Stephen Hawking

It was not an easy process, however, and Wilmut took some 277 attempts to produce Dolly. She spent her life at the Roslin Institute and produced six lambs of her own, including a set of triplets. However, after developing lung disease and arthritis, she was euthanized at the age of six – sheep of her breed (Finn Dorset) have a normal life expectancy of up to twelve years. In the years following Dolly's birth, more mammals were successfully cloned, including pigs, goats, horses and mules. Attempts to clone primates have proved tricky, however, with few surviving beyond the blastocyst stage.

THERAPEUTIC CLONING

In principle, we have the capability to clone human beings, but this practice is banned in many countries. Therapeutic cloning is sanctioned in the UK and America under strict guidelines. Here, cloned embryos are used to extract cells that are genetically identical to the patient. These are harvested at an early stage before they have differentiated into tissue types. Such 'stem cells' can then be manipulated in order to create the kinds of cell required by the patient. One of the great advantages of this practice is that, being genetically identical to the patient, their immune system will not reject the new cells.

Cord blood taken just after birth contains stem cells called haematopoietic blood cells (HSCs) that have the potential to make red and white blood cells

1978	1982	1996
Human insulin is first produced by genetically modified *E. coli* bacteria	The first GM crop is produced, an antibiotic-resistant tobacco plant	Dolly the sheep is born – the first large cloned mammal

Stem cells – embryo vs adult

Use of embryonic stem cells will always court controversy, and for years scientists have been analysing adult stem cells to see if they could offer similar potential. Embryonic stem cells can develop into almost any cell in the body; however, adult stem cells are less versatile.

In 2014, adult human skin cells were turned into stem cells for the very first time using the SCNT technique. It was previously thought impossible to do this because adult stem cells mutate with age. However, researchers in California created stem cells separately from the skin of two men, one of whom was aged 75. This has led to speculation over whether body parts could be regenerated even in old age. Adult stem cell development could lead to tissue transplants to combat a number of serious medical problems, including spinal cord injuries, multiple sclerosis and Parkinson's disease.

and platelets. HSCs are used to treat cancerous blood disorders such as leukaemia in children. Use of cord blood bypasses the ethical debate over use of embryos – since no cloning is required to make the stem cells in this case.

GENETIC MODIFICATION

In 1972, American biologist Paul Berg had developed a technique for splicing DNA into chunks that could be recombined and moved to another organism, even across species. This became known as genetic modification (GM) and it has had a massive impact on the growth of agricultural crops, although like cloning it continues to be controversial. Altering the pattern of genetic code enables possibilities for change in an organism. This has had a huge impact on agriculture, where GM crops are customized in order to have greater pesticide resistance, higher nutritional value and a longer shelf life. For example, identifying the gene that prevents Arctic fish from freezing in the water and inserting it into a crop species can lead to a frost-resistant yield.

In genetic modification, a sequence of nucleotide bases (coding for particular genes) is snipped out of the donor organism's DNA and inserted into the DNA of the organism to be modified.

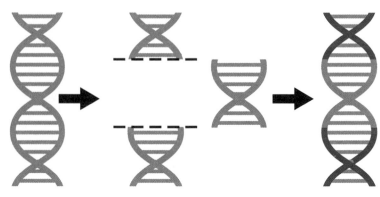

GM isn't just limited to crops, however. In medicine, it is used to create insulin to treat diabetics. This began in 1978, when scientists noticed that if the genetic sequence that codes for human insulin was introduced to the DNA of *E. coli*, the bacteria would grow human insulin. Prior to this, animal insulin was used, which had to be purified. Synthetic human insulin has proved less expensive, is absorbed quicker and causes fewer side effects.

Insects can be genetically modified for a number of reasons, including pest control and crop production. In the Florida Keys, for example, the painful mosquito-borne diseases dengue fever and chikungunya have become a problem. Scientists have bred a GM male mosquito that will mate with wild females to produce offspring that die at the larval stage. A plan to release these insects has met with resistance from residents, who fear the sting of a GM mosquito more than the diseases themselves.

Cloning and GM have fed into an emerging branch of study that was previously the realm of science fiction: synthetic biology. Described as 'genetic engineering on steroids', it aims not only to redesign existing organisms but also to create new life forms entirely.

Feed the world

Golden rice is an intriguing example of how GM crops could solve major world issues. White rice is a staple diet in many countries, but it is not a good source of vitamin A. Millions of children in developing nations are deficient in vitamin A, which can lead to blindness or death.

In 1999, researchers found that if they added two genes to white rice, one from a soil bacterium and another from a daffodil, they stimulated the production of beta-carotene, the pigment that our bodies converts into vitamin A. The researchers called the product golden rice and just one bowl was found to contain 60 per cent of a child's daily requirement of vitamin A. Critics argued that the trials testing golden rice were flawed, while supporters claim it is a viable solution to a huge problem.

The condensed idea
Customizing organisms by editing genes

34 Synthetic biology

Cells are a miracle of engineering – generating energy, building tissue and powering chemical reactions. Replace their DNA with artificial genes and they can become miniature factories, churning out chemicals, including pharmaceuticals and biofuels. This is the dream of synthetic biology.

One of the most innovative areas of modern science, synthetic biology is a leap forward from genetic modification: instead of using DNA from living organisms, it mixes and matches genes created in the laboratory. As a result, new life forms that have never existed in nature can be built. Genetic sequences are designed by computers and manufactured from DNA nucleotide bases that are then implanted into the nuclei of bacterial cells. Unsurprisingly, this scares some people and beguiles others.

In 1972, American scientist Paul Berg became the first person to modify genetic material by cutting DNA from the monkey virus SV40 and splicing it with DNA from a lambda phage virus. Many further experiments were carried out in the 1970s on this so-called 'recombinant' DNA engineering. In the late 20th century, however, scientists began creating custom genes from scratch, which were more cost-efficient than 'editing' existing DNA.

ARTIFICIAL INTELLIGENCE

American scientist Craig Venter is a pioneer in the field of synthetic biology. He established the J. Craig Venter Institute (JCVI) in 2006, with research centres in California and Maryland. In 2010, Venter made headlines across

TIMELINE

1972	1978	2000
Paul Berg is the first person to create recombinant DNA	Human insulin is made from genetically modified *E. coli*	The human genome is mapped, revealing we have 20,500 genes

the world when he declared that his team had created the first synthetic life form, from scratch. It took ten years to achieve, at an estimated cost of $40 million. So how did they do it?

In 2007, the research had reached the stage where JCVI scientists were able to carry out the first-ever whole genome transplant between bacterial cells. They took the DNA of *Mycoplasma mycoides* and successfully transferred it to the nucleus of *M. capricolum*. Venter had his sights firmly set on creating the first synthetic genome, which took another three years to achieve.

WE'RE SPEEDING UP EVOLUTION BY BILLIONS OF YEARS.
Dr Craig Venter

During this time, the team created a computer record of the genetic sequence of the bacterium *M. genitalium*, which they modified to ensure it was not pathogenic. Then they made a real copy of this gene sequence, using just base chemicals, and implanted it into the nucleus of another bacterial cell where it began replicating. The new organism was dubbed *M. laboratorium* and hailed as the first synthetically created life form. Traceable 'watermarks' were inserted into the organism's genome, which indicated that it was synthetic. Critics argued that synthetic biologists were 'playing God' and unleashing potentially dangerous life forms or chemicals into the world. Proponents recognized that modified cells had the potential to become biological factories, churning out proteins, vaccines, drugs and biofuels with many potential benefits.

USES OF SYNTHETIC BIOLOGY

In 2013, researchers at the University of Exeter hit on a method for producing a synthetic fuel similar to diesel. They discovered that a strain of *E. coli* that normally converts sugar into fat could be modified using synthetic biology to convert sugar into fuel. Encouragingly, this new synthetic fuel was found to be compatible with existing modern car engines. The production rate

2007
Craig Venter's team carries out the first whole genome transplant

2010
Venter announces the creation of the first synthetic life form

2013
GM *E. coli* converts sugar into an oil similar to diesel

Genetic Legoland

In 2003, a library of standard synthetic DNA sequences was established, providing Lego-like building blocks that researchers worldwide can use to create new biological systems. Known as BioBricks, some of the most innovative products made from them are showcased at Massachusetts Institute of Technology (MIT) in a yearly competition.

A team from the University of California, Berkeley, designed Bactoblood, an engineered blood substitute made by inserting genes for haemoglobin into *E. coli* cells where the DNA had been destroyed. Bactoblood is claimed to be universally compatible, disease-resistant and cheap to produce. Meanwhile, researchers from the University of Edinburgh engineered bacteria to detect arsenic in water, a serious problem in some nations. If arsenic is present, genes in the bacteria stimulate the production of an acid, which can be detected by a pH test.

is frustratingly slow, however, so the process would have to be ramped up significantly to make this a viable alternative. Other researchers, meanwhile, hope to solve one of the great environmental problems of our time by creating life forms programmed to consume excess atmospheric carbon dioxide. Venter spent two years studying DNA in oceanic blue-green algae, and his researchers have created synthetic photosynthesis cells that absorb CO_2 and light, producing hydrocarbons and oxygen in return.

Nitrogen is key in agricultural crop production, but unlike carbon dioxide, it cannot be assimilated from the atmosphere, despite its abundance. Nitrogen-based chemical fertilizers are needed to boost crop yields and feed our growing population. Their production uses large amounts of energy and when they are added to the soil the greenhouse gas nitrous oxide is released. However, some bacteria can assimilate atmospheric nitrogen because they contain the enzyme nitrogenase. Synthetic biologists are attempting to create crops that contain nitrogenase or have the ability to establish symbiotic relationships with nitrogen-fixing bacteria, perhaps making environmentally damaging fertilizers a thing of the past.

Hazardous waste management is also touted as an area where synthetic biology could have huge impact. Microbes are used to clean-up waste in a process known as bioremediation. In 2010, the Deepwater Horizon disaster discharged 795 million litres (210 million US gallons) of oil across the Gulf of Mexico. Chemical dispersants were dropped on the slick to break it up into droplets that hydrocarbon-chewing microbes could then consume. Synthetic biologists argue that engineered microbes could be used to degrade pesticides, dioxins (chemical contaminants produced by combustion) and even radioactive waste in a cleaner and more cost-effective way.

FIGHTING DISEASE

Human health is another area that could be transformed. Antibiotic resistance has become a pressing problem in recent times, due to overuse of penicillin and related medicines. Researchers are working on engineered bacteriophages – a type of virus that can target specific types of bacteria and destroy them. They do this by invading the bacterial cell, multiplying and causing it to burst open. The viruses also break down the bacterium's protective coat so that it can be recognized by antibiotics or the immune system, allowing them to do their job more effectively. Drugs are also being developed that are programmed to detect cancerous tumours and destroy diseased cells without causing damage to healthy tissue.

Craig Venter (b. 1946)

Born in Salt Lake City, Utah, in 1946, John Craig Venter served in the Vietnam War and his experience during this time inspired him to study medicine and conduct biomedical research. He was awarded his doctorate in physiology and pharmacology at the University of San Diego, California, in 1975. Venter was professor at the State University of New York, Buffalo, and the Roswell Park Cancer Institute.

In 1992, he founded the Institute for Genomic Research, which evolved into the J. Craig Venter Institute, a not-for-profit enterprise where around 250 scientists undertake research to deepen understanding of synthetic and environmental genetics. In 2000, Venter and other scientists announced that they had mapped the human genome, three years before the international Human Genome Project was expected to finish.

Synthetic biology is a young area of science where extensive testing is required, but governments recognize its potential and are prioritizing research and investment. The factories of the future could be microbial, with emissions that save lives and the planet.

The condensed idea
Genetic engineering on steroids

35 Consciousness

Most of us agree we have it, but nobody really understands how it works. Consciousness – how our perceptions, experiences and memories come together in the brain to paint each person's unique picture of reality – is one of the most important issues in neuroscience. And yet it remains one of the most enduring mysteries.

'How can a three-pound mass of jelly that you can hold in your palm imagine angels, contemplate the meaning of infinity and even question its own place in the cosmos?' This quote from neuroscientist Vilayanur Ramachandran encapsulates the mystery of how consciousness can originate from the grey matter in our brains, a conundrum that has baffled scientists, philosophers and psychologists for centuries. Consciousness has been defined as a state of awareness that binds together our perceptions of the world around us into a coherent form and helps us to define our place within it. Those who study consciousness highlight the difference between the brain, as an organ, and the mind, the conscious entity that exists within it and registers feelings and experiences.

Most of us have heard of the French philosopher René Descartes who famously stated *'Cogito ergo sum'* ('I think, therefore I am'). Working in the 17th century, Descartes supported a theory known as 'dualism', which proposed that the mind is a completely separate entity from matter and capable of existing independently. Modern science has disproved dualism through the development of anaesthetics – physical agents that induce unconsciousness in patients.

TIMELINE

1644	1929	1970
René Descartes famously states 'I think, therefore I am'	American philosopher Clarence Irving Lewis coins the term 'qualia'	Self-awareness in different species is gauged in other species using the mirror test

Self-awareness is central to the idea of consciousness. To be self-aware is to realize that you are a conscious being, and to think about your thoughts. For this reason, a number of studies have been carried out to investigate self-awareness in other species.

> CONSCIOUSNESS HAS ALWAYS BEEN THE MOST IMPORTANT TOPIC IN THE PHILOSOPHY OF MIND, AND ONE OF THE MOST IMPORTANT TOPICS IN COGNITIVE SCIENCE AS A WHOLE.
>
> David Chalmers

THE MIRROR TEST

In 1970, psychologist Gordon Gallup devised the mirror self-recognition test to gauge self-awareness in animals. Species are marked with dye then placed before a mirror. If the animal attempts to explore the dye spot after seeing itself in the mirror, it's considered to have recognized its own reflection and to be self-aware. Species that pass the test include humans, primates, elephants, killer whales, bottlenose dolphins, magpies and pigs.

The experiences registered by conscious beings are called qualia, derived from Latin and meaning 'of what sort'. They are subjective, and include sensations like the taste of wine, the scent of a rose or the pain of a headache. They are often difficult to articulate, because one person's perception will differ from other people's. Often, we can only rely on analogies when describing qualia. American philosopher and scientist Daniel Dennett hit the nail on the head when he described them as 'the way things seem to us'. Species that can experience qualia are said to have sentience. This quality is central to the animal rights movement, because it includes the ability to experience pain and suffering.

When we experience qualia in our minds, activity is stimulated in the brain and these processes are called neural correlates of consciousness (NCCs). Many scientific studies have taken place to attempt to link particular brain regions with given emotions or experiences. The development of brain

1996
David Chalmers's influential work *The Conscious Mind* is published

2008
Giulio Tononi puts forward integrated information theory to explain consciousness

2014
American scientists suggest that the claustrum controls consciousness in the brain

Quantum mind

In 1989, two scientists shook up the debate over consciousness by relating it to quantum processes. The eminent Roger Penrose, a professor of mathematics at the University of Oxford, collaborated with anaesthesiologist Stuart Hameroff, and produced a quantum theory of the mind that would become known as 'orchestrated objective reduction' or Orch-OR.

Penrose proposed that consciousness is a result of quantum processes (see page 32) taking place around microtubules in the brain, structures that form the cytoskeleton around which brain cells are built. The duo proposed that these quantum processes account for creativity, innovation and problem-solving abilities. Penrose published his views in the book *The Emperor's New Mind*, which generated a wave of criticism. In 2014, the discovery of quantum vibrations inside these microtubules gave new life to the argument.

imaging techniques, such as electroencephalography (EEG) and magnetic resonance imaging (MRI), has helped enormously.

THE NERVE CENTRE

In 2014, scientists claimed they had discovered a brain region where consciousness can be switched off through electrical stimulation. This is the claustrum, a thin, irregular sheet of neurons located in the centre of the brain. The claim links with a proposal made in 2004 by Francis Crick, one of the scientists who discovered the double helix structure of DNA (see page 129). Crick proposed that to bind together all our experience, human consciousness requires something akin to the conductor of an orchestra. Working with neuroscientist Christof Koch, Crick proposed that this 'conductor' would have to collate information rapidly from various regions of the brain. The pair reckoned that the claustrum was well suited to the task. However, Crick died while working on the idea.

In the 2014 study, a team at the George Washington University in Washington, DC published the results of a study in which they induced unconsciousness in a woman by stimulating her claustrum. The patient suffered from epilepsy and the team were using deep brain probes to try and identify which area of the brain was causing her seizures. The claustrum was one area they investigated. When a current was applied to the claustrum probe, the woman immediately lost consciousness. She stopped what she was doing and didn't respond to any outside stimulus. She regained consciousness the moment the current was switched off, and had no memory of the incident. The test was repeated with the same results. However, this work remains one isolated study – it has not yet been replicated and so is not considered conclusive.

A new attempt to define consciousness is emerging through an area of research called Integrated Information Theory (IIT). It has been pioneered by neuroscientist Giulio Tononi, of the University of Wisconsin-Madison. He believes that consciousness is built not only from our experiences but also from the connections we form between them. If he is right, then scientists attempting to instil consciousness in machines – the field of artificial intelligence (see page 64) – will have their work cut out. For instance, the hard disc on your computer can easily store your lifetime's memories, but it cannot link or integrate them together. A picture of your baby son is distinct from the picture of the toddler he grows into. You can connect the two existentially, but the computer cannot.

The hard problem of consciousness

Why we experience qualia has become known as the 'hard problem of consciousness'. The term was introduced by the Australian philosopher David Chalmers. The 'hard problem', according to Chalmers, lies in the nature of our experiences. He stated, 'When we think and perceive, there is a whir of information-processing, but there is also a subjective aspect.' In other words, why is the sound of a clarinet or the smell of mothballs experienced differently by different individuals?

What Chalmers was looking for was a new approach to explain experience. In the 1990s, when he first proposed the subject, he suggested that the usual explanatory methods of cognitive science and neuroscience were insufficient to elucidate consciousness. The existence of the 'hard problem' has been called into question by some philosophers but Chalmers certainly stirred up the debate.

According to Tononi, the more integrated that concepts and experiences are, the more meaningful they become. He suggests that the amount of integrated information a being possesses indicates their level of consciousness. It's an innovative approach, yet the mystery of how and why we think remains to be solved.

The condensed idea
Mind over matter

36 Language

Complex language is a rich and resourceful gift that is unique to human beings. The source of this ability has long baffled psychologists. In the mid-20th century, the famous linguist Noam Chomsky sparked controversy and debate when he proposed the revolutionary theory that human beings are biologically programmed to acquire language.

Our ability to develop language is remarkable. Other species can communicate via innate skills, or learned systems like birdsong. However, no other creature can express the infinite range of thoughts and ideas that humans generate from a limited range of speech, words and sounds. The reason why has perplexed mankind since the time of Plato.

In the 20th century the development of psychology led to detailed studies of language acquisition. One school of thought proposed that language develops in human beings due to environmental influence and that children acquire language through imitating their parents/carers. This concept is called behaviourism and it was championed by an American psychologist called Burrhus Frederic Skinner. During the 1950s, however, a young American linguist called Noam Chomsky began publishing theories that challenged behaviourism and highlighted its flaws. Chomsky pointed out that language is incredibly complex and that it would take a lifetime for a child to learn its subtle nuances through imitation alone. How could behaviourism account for the fact that most children are fluent in their own language by the age of four or five?

TIMELINE

c.370 BCE	1956	1957
Plato proposes that associating words with meaning is innate	Psychologist Jean Piaget publishes his studies on child cognitive development	Skinner's beliefs on language acquisition are published in the book *Verbal Behaviour*

Through language, children bring forth rich and imaginative concepts when they play and interact with others. Chomsky made an incisive observation when he stated that young children often seem to know more than they have been taught, no matter what environment they have been born into.

LANGUAGE IS A PROCESS OF FREE CREATION; ITS LAWS AND PRINCIPLES ARE FIXED, BUT THE MANNER IN WHICH THE PRINCIPLES OF GENERATION ARE USED IS FREE AND INFINITELY VARIED.
Noam Chomsky

UNIVERSAL GRAMMAR

Chomsky argued that children would never acquire the multitude of skills necessary to understand and formulate language if they were born with only the ability to process what they heard. He proposed that humans have an innate biological capacity for language and that our brains are hardwired to process it. External influence, he argued, can help develop language in children but the ability to master it lies within us. Chomsky's nativist ideas became known as the theory of universal grammar and were truly groundbreaking. He went further to say that humans are born with a 'language acquisition device' in their brains that enables them to encode the key principles of grammar. They build on this foundation by learning vocabulary and syntax in order to construct sentences.

To substantiate his theory, Chomsky made some key points. For instance, languages may vary but children learn their mother tongue in a similar way across cultures. Also, when learning to speak they usually place key words in the correct order. If adults around them make a mistake in syntax, the children are quick to correct them. Children create simple sentences that they could not have acquired from the adults around them, and make common mistakes that cannot be accounted for by behaviourism. For example, it's common to hear infants making errors with words like 'mouses' or 'sheeps', that adults never would.

Chomsky also saw great significance in the fact that language is restricted only to the human species. He highlighted the fact that a child and a kitten are both

1957	1962	2013
Syntactic Structures, Chomsky's germinal work, proposes the basis of universal grammar	Vygotsky's influential work *Thought and Language* is published in the West	Studies show that foetuses can understand rhythm and intonation

Noam Chomsky (b.1928)

Born in Philadelphia, Pennsylvania, Noam Chomsky was the eldest son of teachers William and Elsie. William was an Ashkenazi Jew who fled the Ukraine in 1913. At the age of 16, Chomsky attended the University of Pennsylvania, where he studied philosophy and languages. He was awarded a PhD in linguistics from the University of Pennsylvania in 1955. Chomsky's books on linguistics, published from 1957 onwards, and his theory of universal grammar, made him world-famous and led to a decline in the belief of behaviourism.

He is renowned as a political activist and a supporter of human rights. Following his vocal opposition to US involvement in the Vietnam War he was named on Richard Nixon's 'enemy list'. As one of America's foremost academics, Chomsky has broadened his research interests to artificial intelligence, psychology, computer science and music theory. He remains a leader in his field for establishing what has become known as 'Chomskyan linguistics'.

capable of reasoning and, if they are exposed to the same linguistics, the child will develop language but the kitten never will.

Over the decades, Chomsky's work has become hugely influential, though it has also attracted criticism. His idea that humans have a 'language acquisition device' has not been proved. Also, many linguists have argued that language could develop at an accelerated rate through the desire to interact with other people and form relationships.

SOCIAL SKILLS

In the later 20th century, a theory based on the importance of social interaction emerged. Psychologists and linguists who support this idea argue that children develop language through a desire to communicate with those around them. They are born with powerful brains that allow them to develop new skills when motivated.

These theorists have become known as social interactionists. The idea is based on the work of the early 20th-century Soviet psychologist Lev Vygotsky. He believed that social learning from adults is the primary driver of development in young children. The environment the child is born into and the internalization of language then stimulates cognitive development.

Vygotsky believed that adults are key in influencing a child's cognitive development because they transmit their understanding of language to their children. Vygotsky's work was promoted in the West during the 1930s by the American psychologist Jerome Bruner, but it would take another 30 years for its influence to become widespread.

Other leading psychologists, such as Jean Piaget, emphasized the importance of interaction with peers, to learn key social skills. There are similarities between the work of Piaget and Vygotsky but differences, too. Piaget believed that development had to precede learning, whereas Vygotsky thought that development and learning worked together in symbiosis.

A field of thought called 'empiricism' emerged in the latter part of the 20th century. Its supporters believe that sensory experiences – such as hearing, seeing and touching – are the primary source of language acquisition. The child is born as a 'blank sheet' without any pre-programmed knowledge or language. Empiricists believe that the child's mind will process sensory stimulus and experience and learn language in a social context.

Language and the developing foetus

'Babies learn language in the womb!' This headline appeared across the world in 2013, following research indicating that babies are born with the ability to recognize familiar sounds and language patterns. Studies were carried out on women in the third trimester of pregnancy. They were given a recording to play which repeated the word 'tatata', interspersed with music. At times, the middle vowel was given a different pitch or vowel sound. After the babies were born, they recognized not only the word but its variations, too. Babies in a control group did not recognize the word. The babies studied were only a month old and it remains unknown whether in utero learning has any effect on later development.

Many linguists believe that language acquisition is down to a combination of these nature and nurture approaches. As human beings, we possess an innate ability to master the rules of language, but children develop these skills more effectively via interaction with others. The debate continues, largely thanks to the attention Chomsky's work attracted to this fascinating field of study.

The condensed idea
How we learn to communicate

37 Ice ages

At several times in history, the Earth has become an ice planet. Glaciers stretching from the poles to the equator have left deep scars in the landscape – telltale signs that help scientists date and chart these ice ages. We could even be living in an ice age now, basking in the relative warmth of an interglacial period.

During the last ice age, on average 32 per cent of Earth's land mass and 30 per cent of the oceans were covered in ice. Glaciers locked together to form ice sheets that were 5 kilometres (3 miles) thick in places. Large, hairy animals such as mammoths, sabre-toothed cats, wolves and bears managed to survive the harsh conditions. Humans huddled in shelters built from mammoth bones and wore sewn animal furs.

Our last ice age began 110,000 years ago and ended around 12,000 years ago. Before that, Earth is known to have passed through a number of periods in which it has almost entirely frozen over. Glacial deposits that originated in Greenland and northern Scandinavia have been found at tropical latitudes, hinting that the polar ice caps expanded to meet at the equator.

Modern geological study of ice cores (samples taken from ice sheets), landforms, deep-sea sediments and fossils have all helped to chart the ice ages. Their exact cause remains a mystery, however. Some scientists believed they were caused by periodic changes in Earth's orbit, which influenced solar radiation. Modern theories suggest that varying levels of atmospheric carbon dioxide led Earth into the freezer. Alternatively, the ice

TIMELINE

1741	1837	1830s
Pierre Martel deduces that glaciers change shape over time	Agassiz proposes that Earth was once covered in ice	Karl Friedrich Schimper defines the term *Eiszeit* or 'ice age'

ages could have been precipitated by plate tectonic movements (see page 152) shifting continents and blocking the flow of warm water from the equator.

EVIDENCE OF GLACIATION

Earth's ice cathedrals show first-hand how glaciers leave their mark on terrain. The mighty Mer de Glace (Sea of Ice), on the slopes of Mont Blanc in the Alps, is 7 kilometres (4.3 miles) long and 200 metres (660 ft) deep. The glacier ebbs and flows all the time, swelling, oozing, melting and expanding. Its speed of change is astonishing, with sections of the upper glacier moving up to 120 metres (400 ft) per year. Glaciers are extremely heavy – 1 cubic metre (35 cu ft) of ice weighs 920 kilograms (2,020 lb) – so as they move they carve out huge crevices in the land.

Wobbling planet

'In the mid-20th century, the Serbian geophysicist and astronomer Milutin Milankovitch analysed the obliquity or tilt of Earth's axis of rotation, the wobbling movement of this axis and the shape of Earth's orbit around the Sun. These factors govern the amount of solar radiation hitting the planet, leading to temperature variations.

Milankovitch calculated how these factors would affect radiation levels over set periods. In the 1970s, studies of deep-sea cores proved that glacial periods occurred exactly when Milankovitch predicted. These are now known as Milankovitch cycles. While they may influence interglacial periods, scientists doubt their ability to tip the planet in and out of ice ages.

Most valleys between mountains are v-shaped, but glacial movement creates a trough or u-shape. Evidence of this can be seen around Mont Blanc, the Norwegian fjords and Yosemite National Park in America. As glaciers move, they transport massive rocks and debris that can be deposited kilometres away. Large boulders found in unlikely places, or where their structure differs distinctly to the native rock, are known as glacial erratics.

Moraines offer another important clue. These are mounds of rocky debris made up of till or larger rocks torn from the valley wall as the

1941
Milankovitch publishes his theory on Earth's orbit influencing polar ice caps

1964
W. Brian Harland finds Greenland glacial sediment deposited at tropical latitudes

2014
Scientists use krypton dating to calculate the age of Antarctic ice

The little ice age

From around 1300 to 1870, parts of Earth fell into a 'little ice age'. Mean temperatures were lower than average in the northern hemisphere and winters were extremely harsh. In 1814, London's River Thames froze solid and in 1816 the atmospheric temperature was so cold that birds froze mid-air in Europe and North America. The little ice age was not a global phenomenon, however, and temperatures remained stable in some regions.

The cause of the little ice age is unknown, although some scientists point to volcanic activity. When a volcano erupts it catapults ash and gas into the stratosphere, which reflect solar radiation back into space, cooling the planet below. Radiocarbon dating of plant matter in Canada has revealed that many plants died off during early medieval times, indicating a major cooling. This coincided with a time of increased volcanic activity on Earth.

However, from 1645 to 1715, the Sun's surface also ceased its heat-releasing magnetic storms (an event known as a Maunder Minimum). Some scientists predict that in the 2030s changes in solar activity may force us into another 'little ice age'.

glacier smashed its way through. Lateral moraines form to the side of the glacier and terminal moraines pile up at the endpoint of its journey.

ICE AGE DETECTION

During the mid-18th century scientists began to analyse such clues in the landscape, in particular 'orphan' lumps of rock deposited in valleys where they didn't seem to belong. In the 1830s, German botanist Karl Friedrich Schimper studied mosses that grew on boulders in the uplands of the Bavarian Alps and wondered where the rocks had come from. He visited Bavaria to study them in more detail and concluded that the boulders must have been transported via glaciers. Schimper also proposed the radical idea that at one time Europe, North America and Asia were covered in a vast ice sheet.

Schimper collaborated with his friend the Swiss-American glaciologist Louis Agassiz, who carried out extensive fieldwork in the Alps. They deduced that there must have been global periods of 'obliteration' when Switzerland was covered in a massive ice sheet. In 1837, Agassiz published his theory and is often credited as the first person to propose scientifically that Earth spent periods of its history as a frozen planet. Schimper didn't publish and lost out on the glory, although he did coin the term *Eiszeit*, or 'ice age'.

Reception was unenthusiastic initially, with many contemporaries sticking to the belief that Earth had cooled gradually since early times when it was a volcanic, fiery mass. It took another 30 years before ice age theory would be widely accepted.

DATING ICE AGES

Geologists date the ice ages in a number of ways, through looking at rock scratching or scouring, valley cutting and glacial erratics. Fossil distribution also provides important clues. Over the course of an ice age, organisms that thrived in warmer conditions became extinct, or migrated to lower latitudes where the environment was not so harsh. Further evidence comes from chemical testing, which indicates temperature variations in Earth's history.

Ice ages could last for millions of years. The first began 2.4 billion years ago and four more frozen epochs followed. The most severe was 'Snowball Earth', when the planet is believed to have been almost entirely frozen over. This is thought to have taken place 600 million to 850 million years ago and may have been caused by reduced volcanic activity, which lowered atmospheric levels of the greenhouse gas CO_2 and, therefore, Earth's temperature. CO_2 emissions from volcanoes may ultimately have saved Earth from a permanently icy fate. Normally, CO_2 levels in air and the oceans are balanced but if the oceans were frozen over, then even a gradual increase in CO_2 levels in the atmosphere would warm the planet.

Some scientists believe that the last ice age didn't end at all and that we are currently enjoying an 'interglacial' phase, a more temperate period within the overall course of an ice age. Even if this is true, however, it would take thousands of years for Earth to descend into permafrost again.

THE ROCKS HAVE A HISTORY; GRAY AND WEATHERWORN, THEY ARE VETERANS OF MANY BATTLES; THEY HAVE MOST OF THEM MARCHED IN THE RANKS OF VAST STONE BRIGADES DURING THE ICE AGE.

John Burroughs

The condensed idea
For millions of years, Earth was a frozen planet

38 Plate tectonics

The discovery that Earth's crust contains massive rocky plates that are constantly on the move has explained many of the puzzles of geology, from continental drift, to the formation of deep-sea ridges and the cause of earthquakes. The process is driven by molten matter rising up from Earth's interior.

At the turn of the 20th century, many geologists still believed that Earth's continents were immutable – that they had always existed in their present form and always would do. A number of scientists were beginning to question this mode of thought, however.

Among them was the German geophysicist and meteorologist Alfred Wegener. He couldn't fail to notice that the contours of the Atlantic coastlines of western Africa and South America suggested that they had once been locked together. Wegener speculated that at one time, around 250 million years ago, all the continents were joined up in a 'supercontinent' he called Pangaea. Wegener further hypothesized that the continents had drifted apart over long periods of geological time. He called this phenomenon 'continental displacement', and palaeontological evidence substantiated his ideas. He found records of identical fossilized organisms and rock formations that for some reason cropped up in both Africa and South America, now thousands of miles apart.

Wegener presented his theories to the German Geological Society in 1912, but was met with a cool response. His contemporaries believed that similar fossils

TIMELINE

1912	1929	1960
Wegener presents his ground-breaking theory of continental drift	Holmes states that the Earth's mantle undergoes thermal convection	Hess proposes that the sea floor is constantly spreading

were spread between continents because they were once linked by 'land bridges' that animals could cross. Many also believed that mountains were formed as a result of the Earth cooling down and shrinking to create wrinkles in the crust. Wegener disputed this theory because mountains only form in certain areas, not across the globe. He believed the Himalayas were created when the land mass that became India collided with Asia.

> **IF THE FIT BETWEEN SOUTH AMERICA AND AFRICA IS NOT GENETIC, SURELY IT IS A DEVICE OF SATAN FOR OUR FRUSTRATION.**
> Chester R. Longwell

DYNAMIC EARTH

However, geologists felt that Wegener lacked the credentials to make such radical claims. Another big problem was his inability to explain how the continents had moved. His theories fell into obscurity and might have stayed there if not for the study of palaeomagnetism, which emerged in the 1950s. Scientists discovered that when rocks form they contain an imprint of the direction of Earth's magnetic field at the time. The deeper they dug into Earth's crust, the older the rock layers became, and studies showed that the orientation of the Earth's magnetic field changed according to the age of the rock. The north and south magnetic poles seem to flip quite suddenly after remaining static for long periods of time, reversing on timescales that vary from tens to hundreds of thousands of years.

Studies of the oceanic floor reinforced this. Prior to the Second World War, it was generally believed that the sea floor was flat and featureless, but the development of sonar techniques revealed huge variations, from massive ridges as long as mountain ranges to rifts as deep as the Grand Canyon. Elsewhere, there are large areas as flat as plains. Scientists including Arthur Holmes and Harry H. Hess argued that ocean ridges were created by molten material oozing out of Earth's interior and cooling to form new crust that pushed apart the sea floor. This idea was confirmed in the 1960s when magnetometer readings from the sea-floor plains revealed a pattern of

1966

Dan McKenzie mathematically models mantle convection

1968

Jason Morgan publishes his landmark book *The Theory of Plate Tectonics*

2004

Plate movement causes the devastating Sumatran tsunami, killing 230,000 people stripes with alternating magnetic alignments, proving that the crust grew older moving away from the mid-ocean ridges.

By the 1960s, then, geophysicists realized that Earth's outer shell, or lithosphere, is made up of massive slabs of rock, called tectonic plates, that fit together like jigsaw pieces or segments on a tortoise's shell. There are seven major tectonic plates on Earth – African, Antarctic, Eurasian, Indo-Australian, North American, Pacific and South American. Among them, smaller plates can be found.

Although they are vast, these massive plates move slowly and continuously at a rate of a few centimetres per year, driven by the churning motion of the mantle beneath them. The boundaries where tectonic plates meet are called fault lines, and manifest themselves in various ways.

FAULT TYPES

Divergent faults occur where plates are drifting apart and these create slowly expanding rifts on land, such as the Rio Grande in New Mexico, and deep-sea ridges. Transform faults occur where tectonic plates slide past each other and this grating motion can cause earthquakes. The San Andreas Fault, in California, is a famous example. In 1906, San Francisco was hit by an earthquake that led to the deaths of 600 people.

Earth has seven major tectonic plates and several smaller ones between, jostling for position. Movement of the plates has created deep-ocean ridges, volcanoes and mountain ranges.

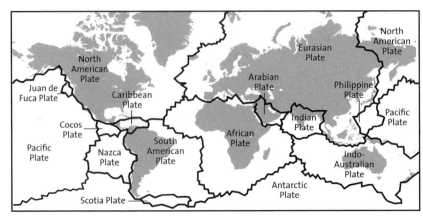

Areas where the plates come together are called convergent faults. Often, the thinner plate of oceanic crust moves beneath another and sinks into Earth's hot mantle forming a so-called subduction zone. Volcanoes and earthquakes are often common in these zones – for example, around the Pacific Ocean's 'ring of fire'. The devastating Sumatran tsunami of 2004 was caused by the Indo-Australian plate subducting below the Eurasian plate. This caused an earthquake of such magnitude that it was estimated to have released the same energy as 23,000 Hiroshima bombs. The ocean was displaced along a 1,000-kilometre (600-mile) rupture, sending waves almost 15 metres (50 ft) high towards the coast of many countries. Around 230,000 people lost their lives.

Measuring motion

The advent of GPS navigation allows us to measure tectonic plate movement with the precision of a few millimetres. GPS receivers on the ground capture signals beamed from a constellation of 30 satellites around Earth. This enables geophysicists to analyse gradual plate movement and the displacement of Earth's crust following an earthquake.

GPS is also used to monitor 'slow-slip' movement in subduction zones where episodic shifting of tectonic plates takes place to release energy. Geophysicists believe that slow-slip indicates a build-up of tectonic stress that will be relieved in future earthquakes. GPS could become a way of highlighting such hotspots before they reveal themselves, with potentially deadly consequences.

Another type of convergent fault involves collisions between thick continental land masses, in which case neither plate can be subducted. Wegener's belief that the Himalayas formed when India slammed into Asia was ultimately proved right: as the plates converged around 60 million years ago, sedimentary rocks were forced together and driven upwards to build the mountain range. The tectonic plates are still jostling against each other, which is why the Himalayas continue to rise gradually each year.

The condensed idea
Earth's surface is always on the move

39 Mass extinctions

Over the past 450 million years, there have been five known events where huge numbers of species have been wiped out from our planet. These are known as mass extinctions and, if climate change continues at its current rate, some scientists believe that we are already descending into another.

By definition, a mass extinction is a global devastation of life within a short period of geological time. This might span a few million years, but that isn't long in Earth's 4.5-billion-year history. There are five known mass extinctions, each marking a transition between geological ages. Theories regarding their cause range from asteroids hitting Earth to climate change and unusual volcanic activity.

These catastrophic events left fossil clues behind in Earth's rock layers. During the 17th century, the discovery and study of fossils began and some of the artefacts recovered were a total enigma. Large bones, horns and teeth were said to be the remains of mythical creatures such as giants, dragons or unicorns. The pioneering French zoologist Georges Cuvier (see box) analysed fossil records in depth and proposed that there were periods in Earth's history when a wide range of species had been wiped out. After British scientist Richard Owen coined the term 'dinosaur' (meaning 'terrible lizard') in 1842, a race began to find as many relics of these creatures as possible.

The oldest fossils were found in the deepest rock layers and corresponded to Earth's earliest living organisms. More complex life forms, such as mammals,

TIMELINE

1796	1842	1962
Cuvier presents research suggesting that some species became extinct	British scientist Richard Owen coins the term 'dinosaur'	Norman D. Newell highlights the evolutionary importance of mass extinctions

reptiles and birds, were found in layers closer to the surface. Rock dating allowed scientists to pinpoint the eras in Earth's history when mass extinctions occurred – namely, where fossils of many species abruptly disappeared. The point at which the dinosaurs were wiped out, for example, has become known as the Cretaceous-Palaeogene (K-Pg) extinction (formerly the K-T extinction). This is because the event separates the Cretaceous and Palaeogene (or Tertiary) periods geologically. The K-Pg boundary is a 1-centimetre (0.4-in) layer of dark clay, sandwiched between dark Palaeogene rock above and lighter coloured Cretaceous rock below. No dinosaur fossils are found above it.

DINOSAUR DEMISE

In the 1970s, American geologist Walter Alvarez was studying rock strata at a gorge in Italy and showed his father, the Nobel Prize-winning physicist Luis Alvarez, the layer

Georges Cuvier and catastrophism

In the 18th century many naturalists resisted the emerging idea that species could become extinct. If that was the case, they argued, eventually all life would die out on the planet. In the 1790s, the famed zoologist Georges Cuvier began presenting sound evidence on extinction. He identified that modern elephants and mammoths were distinct species – meaning that the disappearance of mammoths 4,000 years ago was indeed an example of an extinction.

While examining rock layers in the Paris Basin, Cuvier noticed something intriguing. Successive layers of rock contained fossil layers that would end abruptly, only to be replaced by different types of fossils in the layers above. Cuvier proposed that these gaps in the living record were caused by mass extinction events, a theory that he called catastrophism.

in the rock that indicated when the dinosaurs became extinct. The event was still inexplicable and Luis decided to take some of the clay with him to the lab for analysis. He was astonished at the results. It contained an unusually high concentration of the element iridium – 30 times the level typically found in the Earth's crust. Iridium is rare in Earth's crust, but high quantities of it are found in meteorites and asteroids. A light scattering of iridium drifts down onto our planet from micrometeorites, but there was no way this could account for the quantity seen in the K-Pg layer.

1980
The Alvarezes propose that an asteroid impact destroyed the dinosaurs

1982
Jack Sepkoski and David M. Raup suggest mass extinctions occur at regular intervals

2015
Scientists confirm that climate change is triggering a sixth mass extinction

EXTINCTION IS THE RULE; SURVIVAL IS THE EXCEPTION.

Carl Sagan

Luis and Walter were immediately convinced that the only explanation could be a large asteroid crashing into Earth, destroying the dinosaurs and many other species. They announced their theory in 1980, prompting heated scientific debate. Convincing evidence came in 1990, when surveys of Earth's magnetic and gravitational fields highlighted the rim of a crater in Mexico dating to the era of the K-Pg extinction. Centred around the town of Chicxulub on the Yucatán Peninsula, the crater was 180 kilometres (110 miles) across, meaning that the meteorite causing it must have been at least 10 kilometres (6 miles) across.

It's hard to imagine the devastation caused by the meteorite's impact. It generated a massive earthquake, a mega tsunami, winds of 640 kilometres per hour (400 mph) and a huge fireball that boiled the seas. The dinosaurs that survived the impact would later perish as soot filled the atmosphere, blocking the Sun for months. Vegetation died off and food chains collapsed. It's estimated that the planet took 1 million to 2 million years to recover.

THE GREAT DYING

An asteroid impact has become accepted as the most plausible explanation for the demise of the dinosaurs. This was not, however, the most devastating of Earth's mass extinction events. The Permian-Triassic extinction took place 252 million years ago and destroyed a staggering 96 per cent of marine species and 70 per cent of terrestrial species. All species on Earth today are descended from the survivors. But instead of one devastating event causing this catastrophe, here there is evidence to suggest distinct phases of extinction. An early period of gradual change in the environment seems to have been followed by a sudden event, such as an asteroid impact, volcanic eruption or massive fire. An increase in the microbial population that releases methane into the atmosphere could also have triggered a greenhouse effect, sending the planet into chaos.

Some scientists believe that the assembly of supercontinent Pangaea (see page 152) may be the root cause. They reckon that as continental plates moved towards each other, marine habitats were destroyed and ocean currents disrupted, which imbalanced regional climates.

Earth's five main mass extinctions

In the 20th century, scientists identified five major extinctions in Earth's history. The first took place in the late Ordovician period, 438 million years ago (mya). Most life was still in the sea, so marine creatures such as brachiopods and trilobites were drastically reduced in number. The next mass extinction was the late Devonian (360 mya) and this wiped out 30 per cent of species on Earth. Coral reefs were particularly affected and would not flourish again for millions of years.

The biggest mass extinction of all, known as 'the Great Dying' was the Permian-Triassic (252 mya), in which 96 per cent of marine species were wiped out. Two or three phases of extinction led to the Triassic-Jurassic (201 mya) event. Finally the Cretaceous-Palaeogene event (66 mya) killed off the dinosaurs and up to 70 per cent of Earth's other species.

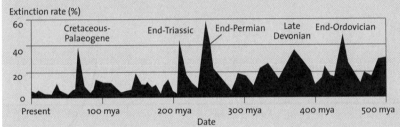

Earth's five known mass extinction events, showing the percentage of species wiped out each time. Note, this is the percentage of species that leave fossils – the percentage of all may be much higher.

If climate change continues unabated, we could be on course for a sixth mass extinction. Even conservative projections indicate that 16 per cent of Earth's species will die out before the end of the 21st century, with vertebrates vanishing at a higher rate than during any previous mass extinction. The European Space Agency is developing technology to deflect asteroids in space before they ever reach Earth. It would seem, however, that a greater threat lies much closer to home.

The condensed idea
Life on Earth is periodically ravaged by catastrophic events

40 Climate change

Climate change caused by rising carbon dioxide levels in Earth's atmosphere was first proposed by the French scientist Joseph Fourier in 1824. It would take another 150 years before the dangers of man's impact on the planet would begin to be recognized and steps taken to reduce our collective 'carbon footprint'.

Climate change is the greatest threat to Earth's future, and many climate scientists believe that we are approaching a 'tipping point', after which little can be done to prevent dire consequences. It's a dystopian vision of the future: sea levels rise to the point where cities such as London and New York are flooded; mass migration from uninhabitable countries causes chaos and warfare; vast swathes of plant and animal life die off, leading to famine on a biblical scale.

The huge population growth of the last 100 years has exacted an enormous toll on the planet through disturbing its delicate climatic balance. Earth enjoys a mild and temperate climate thanks to the atmosphere – the thin layer of gases that cloaks the planet and prevents it from becoming scorching hot or freezing cold, like others in our Solar System. Sunlight passes through our atmosphere as visible light and warms the ground below. Earth's surface emits energy in the form of infrared thermal radiation, which has a longer wavelength than visible light. It is harder for this form of energy to pass through Earth's atmosphere and it gets absorbed by greenhouse gases such as carbon dioxide (CO_2), methane and ozone – which then re-radiate some of it back down towards the ground. Although

TIMELINE

1824	1896	1938
Fourier hypothesizes that Earth's atmosphere somehow warms the planet	Arrhenius models how CO_2 influences the temperature of our planet	Callendar illustrates a rise in Earth's temperature over the previous 50 years

called the 'greenhouse effect', the process is slightly different to the action of a greenhouse, which works by preventing warm air from rising up and escaping.

EARLY THEORIES

The greenhouse effect has been well publicized over the last 40 years, but it is not an entirely modern theory. In 1824, the French scientist Joseph Fourier calculated that given Earth's distance from the Sun, the planet should have a much colder surface temperature. He proposed that somehow the atmosphere insulated Earth, keeping heat in. For this reason, Fourier is credited as being the first person to identify the greenhouse effect – even though no one knew how it actually happened.

In the following decades, however, scientists came to understand that CO_2 absorbs infrared radiation and, in 1896, Swedish chemist Svante Arrhenius constructed a mathematical model to show how CO_2 influenced temperatures on Earth. He established that if CO_2 increased or decreased then the temperature would, respectively, increase or decrease as well. Arrhenius calculated that if CO_2 levels doubled in our atmosphere, Earth's temperature would increase by about 4°C (7°F). If CO_2 levels halved, we would enter another ice age.

Arrhenius was the first to recognize that the burning of fossil fuels, vastly accelerated by industrialization, would increase atmospheric CO_2 and warm the planet. However, his hypothesis didn't provoke dire predictions for Earth's future. At the time, it seemed implausible that human activity could significantly influence Earth's temperature. Scientists believed that excess CO_2 would be absorbed in the oceans. Arrhenius proposed that increased CO_2 could make conditions 'more equable', boosting plant growth and food production. Some even argued that it could help stave off another ice age. Nevertheless, Arrhenius's findings were disputed and fell into obscurity for almost 50 years.

CLIMATE CHANGE IS NOT JUST A PROBLEM FOR THE FUTURE. IT IS IMPACTING US EVERY DAY, EVERYWHERE.
Dr Vandana Shiva

1961
The Keeling curve demonstrates a steady increase in atmospheric CO_2

1989
The Montreal Protocol to protect the ozone layer is ratified

2014
The United Nations warns of limited time to reverse climate change

The direct correlation between increasing CO_2 levels in the atmosphere (measured in parts per million or ppm) and rising global temperatures since the late 19th century. The line traces atmospheric CO_2 while the bars denote temperature. The two are clearly correlated.

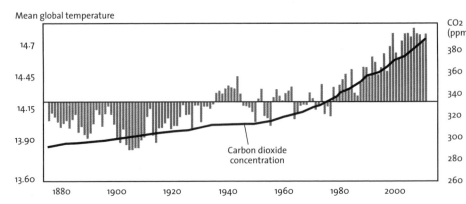

PROOF OF GLOBAL WARMING

In 1938, British engineer Guy Callendar tried to revive Arrhenius's theory. He collated temperature measurements from the 19th century and compared them with atmospheric CO_2 levels. Callendar deduced that Earth's temperature had risen by 0.5°C (0.9°F) over the preceding 50 years and proposed that increased CO_2 was the cause. His handwritten calculations have since been proved to be accurate, and the influence of CO_2 on global temperatures is known as the Callendar effect.

In the 1950s, it became evident that the excess CO_2 released by the burning of fossil fuels was not being absorbed by the oceans. Scientists were starting to understand that increased CO_2 could be dangerous to the planet. In 1961, the American geochemist Charles Keeling illustrated that CO_2 levels were rising steadily on what has become known as the Keeling curve. Arrhenius's calculations were subsequently resurrected and refined through computer technology.

The message was getting through from scientists to governments – the planet was warming quickly and hurtling towards a devastating short-term change. In 1965, American president Lyndon B. Johnson delivered a special message to Congress that stated: 'Air pollution is no longer confined to isolated places. This generation has altered the composition of the atmosphere on a global scale through radioactive materials and a steady increase in carbon dioxide through the burning of fossil fuels.'

HEAT RISES

Scientists can calculate historical mean temperatures on Earth by analysing bubbles of air contained in glacial ice. As Earth moved out of the ice ages, the average global temperature rose by a total of 4–7°C (7–13°F) over 5,000 years. In the last century, Earth's temperature has risen by 0.7°C (1.3°F) This is around ten times faster than the average rate of temperature rise following the last ice age. In 2015, sea levels were found to be rising at a significantly higher rate than previously predicted. Since 1990, the oceans have risen by 3 millimetres (0.1 in), which is 25 per cent higher than original estimates.

According to a United Nations report published in 2014, we only have a few years left to avoid a catastrophic warming that would render currently populated parts of the planet uninhabitable. Those include Northern America and Europe. A temperature increase above 2°C (3.6°F) is acknowledged to be the 'tipping point'. The time to act is now.

The Albedo effect

The Arctic is warming almost twice as rapidly as the rest of the planet, which has resulted in a 14 per cent decrease in its sea ice since the 1970s. This is influencing a key regulatory process called the albedo effect (taken from the Latin work meaning 'whiteness'). A planet's albedo is defined as the fraction of the solar energy falling on its surface that is reflected back into space. The albedo is given as a number between 1 and 0 – snow-covered areas have an albedo close to 1 (100 per cent of light is reflected back), whereas darker surfaces have an albedo closer to zero. Earth's overall albedo is 0.35, but as the polar icecaps melt due to global warming, the albedo decreases – meaning less solar energy is reflected back into space. This warms the planet even more, causing more ice to melt, and lowering the albedo still further in a runaway feedback loop.

The condensed idea
Global warming is the greatest threat to Earth

41 The Copernican Solar System

The heliocentric view of our Solar System was first proposed by the 16th-century astronomer Nicolaus Copernicus, who overturned the long-held misconception that Earth was at the centre of the universe. He gained widespread condemnation by stating such a radical idea and it was only published as he lay on his deathbed.

The theories of Nicolaus Copernicus ushered in a new era in our understanding of the universe. For thousands of years even the most learned scholars believed that the Sun and the planets of the Solar System orbited Earth and that our planet was stationary at the centre of the universe. This geocentric notion was proposed by the ancient Greek philosophers, including Plato and Aristotle.

The fourth-century BCE mathematician Eudoxus of Cnidus created a model illustrating how celestial bodies could orbit Earth. He believed that they were positioned on spheres that encircled Earth at increasing distances, with the Moon on the closest sphere, followed by Mercury, Venus, the Sun, Mars, Jupiter, Saturn and finally the distant stars. Aristotle, who developed the idea, believed that each planet was moved by its own god. Aristarchus of Samos, in the third century BCE, was a lone voice questioning the geocentric view. He believed that Earth rotated on its axis and revolved around the Sun. But his theory was rejected under the weight of Aristotle's ideas.

TIMELINE

4TH CENTURY BCE	3RD CENTURY BCE	1543
Astronomers believe that the planets and the Sun orbit Earth	Aristarchus of Samos proposes a heliocentric view of the Solar System	Copernicus publishes his vision that the planets orbit the Sun

However, there were some very obvious anomalies in the geocentric model that baffled early astronomers. For example, it didn't explain the varying brightness of the planets, or the fact that periodically they appear to change direction, literally moving backwards across the sky.

> TO KNOW THAT WE KNOW WHAT WE KNOW, AND TO KNOW THAT WE DO NOT KNOW WHAT WE DO NOT KNOW, THAT IS TRUE KNOWLEDGE.
> Nicolaus Copernicus

Egyptian astronomer Ptolemy of Alexandria devised a convenient theory to resolve this problem. In his *Almagest* (*Great Compilation*) of 150 CE, he proposed that as each planet travelled along its sphere, it also completed smaller circular motions called 'epicycles'. When the planet's epicycle corresponded with the movement of its sphere it was said to be in 'prograde' motion across the sky. If the epicycle was opposed to the sphere's direction it was in 'retrograde' motion, appearing to slow down and reverse. Early astronomers believed this could also account for variations in planetary brightness.

REVOLUTIONARY THEORY

Astronomers trusted Ptolemy's theory for another 1,400 years, until the brave Polish astronomer Nicolaus Copernicus turned the entire model on its head. Copernicus studied astronomy at the University of Kraków and was aware of the work of Aristarchus of Samos. He questioned the geocentric theory of epicycles, believing it to be plagued with flaws and logical holes, and began instead to develop his own model.

Following years of observations and calculations, Copernicus established a heliocentric view of the Solar System with the Sun at its centre instead of Earth. In his model, all of the planets orbited the Sun, doing so in the order that we now know to be correct – Mercury, Venus, Earth, Mars, Jupiter and Saturn (Neptune and Uranus were discovered later). This 'new' vision of our Solar System explained why Mercury and Venus remain close to the Sun as seen from Earth.

1610	1609–19	1687
Galileo reinforces Copernican theory through telescopic observations of Venus	Kepler defines the laws of planetary motion and elliptical orbits	Newton reveals that gravity holds planets in orbit around the Sun

Nicolaus Copernicus (1473–1543)

Copernicus was born in the Polish town of Toruń on 19 February 1473. His father, Nicolaus, was a successful merchant and his mother, Barbara Watzenrode, came from a leading mercantile family. Following his father's death, the young Copernicus was taken under the wing of his uncle Lucas Watzenrode, who became bishop of Warmia. He supervised Copernicus's education and guided him towards a career in the Church.

After studying a broad range of subjects, including astronomy, at Kraków University, Copernicus attended the University of Bologna. This was a key episode in his career because he boarded with the principal astronomer of the university, Domenico Maria de Novara. In 1503, Copernicus was awarded a doctorate in canon law and he returned to Poland to work for the Catholic Church in an administrative role, collecting rents, overseeing finance and providing medical care.

His reputation as an astronomer grew, and in 1514 he was one of the experts consulted by the Church on the revision of the calendar, which had fallen out of alignment with the Sun. At this time, he also completed a 40-page manuscript called *Commentariolus* (*Little Commentary*) outlining initial ideas on his heliocentric theory. This was a precursor to the more detailed work, *De Revolutionibus Orbium Coelestium*, which he resisted publishing for many years, fearing opposition and scorn from the establishment.

Copernicus fell ill after completing this manuscript and died on 24 May 1543. Legend has it that he awoke from a coma to find his book in his hand, looked through it and passed away peacefully. He is buried in Frombork Cathedral, where he worked from 1522 until his death.

Copernicus also proposed, correctly, that the Earth rotates daily on its axis and completes an orbit of the Sun every year. His theory dispensed with the anomalies of Ptolemy's geocentric vision, in particular the epicycles. Retrograde motion could now be explained naturally because Earth is not stationary, but is moving in its orbit all the time. As a result, when it laps slower-moving outer planets, such as Jupiter, the gas giant appears to reverse direction in the sky. This is essentially the same optical illusion you experience in a car that overtakes a slower-moving vehicle.

MAGNUM OPUS

Copernicus published his theories in 1543 in a work entitled *De Revolutionibus Orbium Coelestium* (*On the Revolutions of the Heavenly Spheres*). He knew that his ideas would spark controversy because they challenged the teachings of the Roman Catholic Church, and indeed the work remained on the Catholic Church's forbidden reading list until 1822. Ironically, Copernicus himself was a devout Catholic and didn't believe that his vision of the universe contradicted the Bible. He died of a cerebral haemorrhage just a few months after his magnum opus was published.

However, Copernicus's theories were not without their flaws. He still believed that the planets orbited the Sun in perfect circular motion, and was forced to retain some of Ptolemy's epicycles in order to make his heliocentric model as accurate as the one it sought to replace. Partly because of this, few astronomers accepted Copernicus's theory at the time.

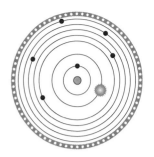

In fact, circular motion was only abandoned in the early 17th century thanks to the work of German astronomer Johannes Kepler, who correctly deduced that the planets actually move around the Sun in elliptical orbits. This crucial addition finally removed the need for epicycles. Around the same time in Italy, Galileo observed four moons orbiting the planet Jupiter with an early telescope, showing once and for all that not everything rotated around the Earth. He also studied Venus, noting phases (similar to those of the Moon) that could only be explained if it travelled around the Sun. Galileo was put under house arrest for his 'unorthodox' beliefs, but the geocentric revolution now picked up speed. In 1687, Isaac Newton put the final nail in the coffin of geocentrism when he discovered the force that holds the planets in their elliptical orbits around the Sun – gravity.

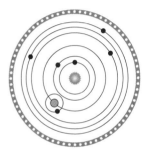

In the ancient geocentric view of the heavens (top), the Sun and planets orbit the Earth. The Copernican view (below) put the Sun firmly where it should be, at the centre of our Solar System

Copernicus's legacy is inestimable, and there is no doubt that he has earned the moniker 'the father of modern astronomy'. His vision also established the Copernican principle – that Earth does not occupy a privileged place in the universe. As Carl Sagan said, 'We find that we live on an insignificant planet, of a humdrum star lost in a galaxy tucked away in some forgotten corner of a universe in which there are far more galaxies than people.'

The condensed idea
Earth is not the centre of the universe

42 Galaxies

Stars and planets are not evenly spread throughout the universe but are instead clustered into cosmic islands known as galaxies, each separated by vast gulfs of empty space. Our Sun and Solar System are all part of the Milky Way galaxy, while the universe beyond is home to many billions more.

Take a look up on a clear and dark summer's night and you'll see a magnificent band of light arching across the sky. This is the Milky Way. It's our galaxy – a collection of hundreds of billions of stars, spanning almost 120,000 light years. Galaxies come in many shapes and sizes. Ours is a spiral disc, which when viewed from the inside creates the bright band visible across the night sky. The display is at its best in summer because that's when the night side of the Earth faces the bright galactic centre. It took a while for us to realize this much. Early astronomers thought the bright haze of the Milky Way was due to glowing gas. It was Italian scientist Galileo, in the early 17th century, who first turned his telescope on the Milky Way to reveal that it is in fact made of stars. Yet it was still over two centuries before the real meaning of this sank in.

In 1750, English astronomer Thomas Wright published a book called *An Original Theory of the Universe*, in which he advocated the notion that the Milky Way is a disc of stars, within which the Sun and the Solar System are embedded. But he went further – what if, supposed Wright, there are other such groups of stars dotted around the universe? The German philosopher Immanuel Kant read the book and agreed with Wright's conclusions. He called these distant groups

TIMELINE

1750	1760	1860
Thomas Wright suggests that islands of stars litter the universe	Charles Messier detects the first of his mysterious nebulae	Spectroscopy proves that Messier's nebulae are made of stars

of stars 'island universes' and – by analogy with our Solar System – speculated that these faraway systems of stars might be rotating.

PSEUDO COMETS

The theory was still being treated with great scepticism by the scientific community when, in the 1760s, French astronomer Charles Messier began to record faint, fuzzy patches of light in the night sky through his telescope. Messier was a comet hunter, and spent many hours scanning the skies for the hazy patches of light created when a chunk of icy debris from the outer Solar System falls towards the Sun, heats up and develops a gaseous atmosphere or 'coma'. Messier's search for comets was repeatedly complicated by the presence of numerous coma-like fuzzy objects that remained remained static in the night sky (unlike comets, which noticeably changed position as they orbited the Sun).

Between 1760 and 1784, Messier discovered more than 100 of these objects, and drew up a catalogue that listed them, chiefly as a way of speeding up his comet searches. He and other astronomers soon determined that some of the Messier objects were clouds of tightly packed stars, but others remained diffuse at even the highest magnifications and became known as 'nebulae' (after the Latin word for 'cloud').

It wasn't until 1860 that the true nature of the nebulae became clear. The breakthrough came as a result of a new experimental tool in the astronomer's bag, called spectroscopy. This is a way to determine the chemical composition of an astronomical object by analysing its light. It works by splitting the object's light into its rainbow spectrum of colours and then measuring the

THERE ARE AT LEAST AS MANY GALAXIES IN OUR OBSERVABLE UNIVERSE AS THERE ARE STARS IN OUR GALAXY.
Professor Sir Martin Rees

The structure of a typical spiral galaxy, such as the Milky Way.

1922	**1943**	**2012**
Hubble and Humason prove the nebulae are galaxies far beyond the Milky Way	The first active galaxy is discovered by American Carl Seyfert	The Hubble Telescope sees thousands of galaxies in the eXtreme Deep Field

Edwin Powell Hubble (1889–1953)

Edwin Hubble was born in 1889 in Marshfield, Missouri. He studied mathematics and astronomy – a topic that had fascinated him since childhood – at the University of Chicago, graduating in 1910. He also excelled as an athlete in his youth.

Despite Hubble's keen interest in astronomy, his father wanted him to pursue a career in law – a request that he initially acceded to, gaining a place to study the subject at Oxford. However, following his father's death in 1913, Hubble returned to Chicago and obtained a PhD in astronomy at Yerkes Observatory. This led to a permanent post at Mount Wilson Observatory, California, where he remained for the rest of his life.

brightness of each particular colour. Chemical elements and compounds emit and absorb different amounts of each colour, creating a unique pattern of bright and dark bands in the spectrum.

When astronomers applied spectroscopy to Messier's nebulae, the technique revealed there were two distinct types – some were indeed clouds of gas as they appeared, but others showed the combined light of countless stars. But where were these groups of stars located?

NEBULAE UNMASKED

American astronomers Edwin Hubble and Milton Humason provided the answer to that question in the 1920s. Hubble came up with a clever way to determine the distances to the nebulae by measuring the brightness of what are called Cepheid variable stars. The brightness of these stars periodically rises and falls – and, crucially, each Cepheid's average brightness can be inferred from the period of its variations. This means that if you can find a nebula containing a Cepheid variable star, then measuring the period of the star's brightness variations tells you its absolute average brightness. Armed with that information, along with measurements of its average apparent brightness as seen from Earth, you can work out how much its brightness has dimmed with distance, and thus deduce how far away it is.

Hubble and Humason's observations showed that the nebulae were typically millions of light years away. Compared with the size of the Milky Way (determined by American Harlow Shapley in 1918 as around 100,000 light years), it was clear that the nebulae were situated far beyond. Thomas Wright's assertion that they are distant copies of the Milky Way had been proven correct. Consequently, the nebulae were renamed galaxies.

Many galaxies have been seen to be discs of stars with a flattened spiral structure, suggesting rotation, just as Kant suggested. But other types have now been seen, too, including ball-shaped elliptical galaxies and amorphous irregulars with very little discernible structure. Ellipticals are thought to be created by mergers between spirals, and some galaxies have even been caught in the act of spectacular collisions, casting streams of stars and debris into deep space.

OUR NEIGHBOURHOOD

Spiral structure has since been confirmed in the Milky Way. The Sun and its Solar System are believed to orbit around 27,000 light years from the centre of the galaxy, on the inside edge of the so-called Orion spiral arm. Our galaxy is thought to rotate roughly once every 225 million years.

Active galaxies

In 1943, American astronomer Carl Seyfert discovered a number of spiral galaxies that were far brighter than others. They became the first in a new class of astronomical objects, called active galaxies.

Active galaxies have now been discovered in all shapes and sizes. Each is thought to be powered by a supermassive black hole (see page 188) in its centre. The black hole is sucking in material from the surrounding galaxy, and as this falls in, it gets compressed and heated to enormous temperatures, causing it to glow brightly. The very brightest active galaxies are known as quasars. They are the most distant objects in the universe so far detected.

In 2012, NASA released the 'eXtreme Deep Field', a long-exposure image from the Hubble Space Telescope that captures galaxies as they were an astonishing 13.2 billion years ago. It contains some 5,500 galaxies packed into an area of just 2.4 square arc minutes – a little over one hundred-millionth of the entire sky. That suggests there could be hundreds of billions of galaxies out there, each as large and populous as the Milky Way – a humbling thought indeed.

The condensed idea
Islands in the cosmos

43 The Big Bang

The Big Bang theory says that almost 14 billion years ago, our universe was born in a hot, dense fireball – from which it expanded and cooled to form the galaxies and stars that we see today. Deducing this from the scant evidence available took some ingenious cosmic detective work.

Long before the development of science, almost every culture on Earth fostered its own ideas about how the universe began. Most of them were completely wrong. The Mesopotamians, in the 16th century BCE, believed in a flat Earth afloat in a cosmic ocean. The ancient Greeks believed that the Sun, Moon, planets and stars all inhabit crystalline spheres that revolve around the Earth (see page 164).

As science became established in the 18th and 19th centuries, evidence came to light that the Earth and stars were vastly older than previously believed, and that geological and cosmic processes operated over hundreds of millions, even billions of years. Most scientists therefore came to the conclusion that the universe had been here forever, and would continue to do so – a so-called 'steady state' theory. We now know that this is impossible because of the second law of thermodynamics. The law says that the universe's entropy – essentially, how 'well mixed' it is – is constantly increasing (see page 20). If the universe was infinitely old, then its entropy should be infinite, too, and this would make everywhere the same temperature – lacking the distinction between hot, massive stars and cold, empty space that we see today.

TIMELINE

1927	1929	1950
Lemaître applies relativity to the universe and predicts cosmic expansion	Hubble discovers the expansion of the universe, supporting Lemaître's work	Hoyle coins the phrase 'Big Bang' in a derogatory remark about the theory

DYNAMIC COSMOS
Our best theory for the origin and evolution of the universe is the Big Bang model. In 1927, a Belgian priest and physics professor called Georges Lemaître applied Einstein's general theory of relativity (see page 28) to the universe at large. The resulting model predicted that the universe must either be expanding or contracting, depending on the average density of the matter it's filled with – the one thing it can't do is remain eternally unchanging, as required by the steady state theory.

> IN THE BEGINNING THERE WAS NOTHING, WHICH EXPLODED.
>
> Terry Pratchett

From his model, Lemaître realized that if the universe is expanding as a whole, then the rate of expansion between two points must get faster the further apart they are. Two years later, in 1929, American astronomers Edwin Hubble and Milton Humason observed this very effect. They studied distant galaxies, finding that the further away each galaxy was, the more its light became stretched or 'red-shifted' by its motion away from us (see page 181) – proving that the universe is expanding in precise agreement with general relativity.

INFINITE DENSITY
But Lemaître saw another key consequence to his model. Trace the cosmic expansion backwards through time and the galaxies get closer and closer together until there comes a point when they must all have been packed in a tiny dense sphere – and this was the state in which the universe began. Lemaître published his model in 1931, showing how the universe sprang into existence in what we today call the Big Bang (though that name was not coined until 1950, in a mocking

The expanding universe is rather like the surface of an expanding balloon – distant objects move apart faster than those close to each other.

1964
The cosmic microwave background radiation (CMBR) is accidentally discovered

1992
The COBE satellite returns the first images of structure in the CMBR

2015
The Planck space probe pins down the age of the universe to 13.8 billion years

The inflationary universe

During the late 1960s, British mathematicians Roger Penrose and Stephen Hawking showed how general relativity required the initial state of our universe in the Big Bang model to be a point of infinite temperature and density known as a 'singularity'. Similar states are thought to prevail in the hearts of black holes (see page 188), where they generate gravity so intense that not even light can break free. How then did the universe escape from this superdense state, and not remain trapped forever?

The solution is an idea called inflation, proposed by Alan Guth, at the Massachusetts Institute of Technology, in 1980. It says that during the first fractions of a second after the instant of creation the universe underwent a phase of superfast expansion. No sooner had inflation begun than it stopped again, leaving the universe expanding at a more sedate pace, which is how we find it today.

Inflation is motivated by effects known to have taken place in high-energy particle physics, in particular the breaking of particle symmetries as the universe gradually cooled (see page 43). These symmetry-breaking events are thought to trap regions of space in what's called a 'false vacuum', a state that can produce an antigravity force strong enough to overcome the powerful gravity of the singularity.

Inflation also suggests a mechanism for the formation of galaxies and clusters in the universe, namely that quantum mechanical fluctuations (caused by Heisenberg's uncertainty principle – see page 37) are blasted up to cosmic scales by the rapid expansion. Recent measurements by space probes support this idea.

remark made by British astronomer and steady-state proponent Fred Hoyle).

In the 1940s, American cosmologists George Gamow, Robert Herman and Ralph Alpher showed that if the theory was correct then light from the superheated fireball must have been red-shifted just like that from Hubble and Humason's galaxies, leaving a faint microwave echo pervading space. This signal, known as the 'cosmic microwave background radiation' (CMBR), was found in 1964 by American radio astronomers Arno Penzias and Robert Wilson – inadvertently, as it turned out. They were experimenting with an extremely sensitive radio receiver at Bell Labs, New Jersey, and couldn't fathom the source of the faint noise that was plaguing their results. It was only when a group of cosmologists (scientists who study the universe at large), led by Robert Dicke at Princeton University, learned of Penzias and Wilson's work that the true source of the noise became clear. Penzias and Wilson received the 1978 Nobel Prize for their contribution.

Gamow and Alpher also looked at the abundances of chemical elements cooked up by nuclear reactions in the first 20 minutes after the Big Bang. Their finding of 25 per cent helium and 75 per cent hydrogen exactly matches what astronomers have found in the oldest generations of stars (newer stars will have been contaminated with chemical elements formed since).

STRUCTURE FORMATION

The CMBR has since been studied in pin-sharp detail from above the obscuring murk of the Earth's atmosphere. NASA's COBE (Cosmic Background Explorer) satellite was the first to see small fluctuations in the radiation's temperature across the sky. These reveal tiny irregularities in the density of matter shortly after the Big Bang, out of which the galaxies and clusters that populate the universe today ultimately grew through gravity. In the early 21st century, two new space probes – NASA's Wilkinson Microwave Anisotropy Probe (WMAP) and the European Space Agency's Planck experiment – measured the CMBR and its structure in unprecedented detail. Their findings have helped cosmologists to pin down many of the parameters governing the behaviour of our Big Bang universe and yielded the most accurate estimate yet for its age: 13.8 billion years.

BEFORE THE BIG BANG?

Scientists are not yet clear on what caused the Big Bang – or what existed before it, if anything. As it's presently understood, the event created not just the matter and energy filling our universe, but also the very space and time that it's made from. Some advanced ideas in particle physics, such as string theory, offer thoughts on what happened prior to the creation of our universe, but for the time being these are largely untestable. Right now, scientists are more concerned with identifying the strange forms of matter that fill the universe, and understanding its evolution and ultimate fate.

The condensed idea
The universe began in a hot and dense state

44 Dark matter

Despite the best efforts of science, 95 per cent of our universe remains a mystery. This is dark matter – stuff we know must be out there because of its gravitational influence on bright objects, but which we simply cannot directly see. Without dark matter, the cosmos of today could not have existed.

Look up on a dark night and the sight is quite overpowering – millions of stars and galaxies stretched out before you, lighting up the dark like a cosmic metropolis. It can come as something of a shock to then discover that what you're looking at is in fact only 5 per cent of the complete picture. The other 95 per cent is invisible – it's what astronomers call 'dark matter'.

The first serious proposal that there's more to the universe than meets the eye came from the Swiss astronomer Fritz Zwicky during the 1930s. Zwicky, working at the California Institute of Technology, had been studying the Coma Cluster – a group of over 1,000 galaxies, 320 million light years away. The galaxies in the cluster are all moving this way and that, like a swarm of bees. Zwicky measured the typical speeds with which individual galaxies were moving. He could see that the galaxies were not flying off into space, and – like a projectile fired from the surface of a planet that ultimately falls back down – this meant that the cluster's gravity must be strong enough to hold each galaxy in place.

MISSING GRAVITY
But Zwicky was in for a shock. When he totalled up all the bright matter that he could see in the galaxy, and plugged the amount of mass that this

TIMELINE

1933	1962	1983
Fritz Zwicky discovers dark matter in the Coma Cluster	Vera Rubin finds the first evidence for dark matter in galaxy rotation curves	Mordehai Milgrom proposes MOND as an alternative to dark matter

represented into Newton's law of universal gravitation (see page 12), it seemed the cluster's gravity was hundreds of times too weak to rein the galaxies in. Zwicky realized that something else must be generating the necessary gravitational force, and he referred to this hidden mass as 'dark matter'.

Zwicky's work was not accepted at the time – and, indeed, his calculations were a little off beam – but his general conclusions were correct: there is much more mass in the universe than can be seen. It wasn't until the 1970s that astronomers began to take the idea seriously, following American astronomer Vera Rubin's studies of the 'rotation curves' of spiral galaxies.

WE BECAME ASTRONOMERS THINKING WE WERE STUDYING THE UNIVERSE, AND NOW WE LEARN THAT WE ARE JUST STUDYING THE 5 OR 10 PERCENT THAT IS LUMINOUS.
Vera Rubin

IN A SPIN

Spiral galaxies are whirlpool-like collections of stars found in deep space – much like our own Milky Way (see page 168). A rotation curve is essentially a graph, showing how the galaxy's rotation speed changes with distance from its centre. Standard gravitational theory predicts that, in the galaxy's core region, the speed of rotation should increase in direct proportion to distance, before decreasing again in the outer spiral arms.

But that wasn't what Rubin found at all. She saw the graph grow in proportion to radius in the core, as expected, but then flatten off in the spiral arms, rather than decaying away as the theory had predicted. The only way this could happen is if the outer regions of the galaxies contain vast amounts of hidden material, providing extra gravitational force to drive the rotation faster. Rubin did her calculations, and unlike Zwicky her numbers were bang on. She found that the visible stars, gas and dust in a spiral galaxy constituted just 5 per cent of its total mass.

2005	2011	2013
Dark matter is proven essential for forming structure in the universe	CRESST collaboration reports detection of possible dark matter particles	ESA's Planck spacecraft confirms that up to 95 per cent of the universe is invisible.

Alternative theories

Not everyone is convinced that dark matter is necessary. Physicist Mordehai Milgrom, of the Weizmann Institute in Rehovot, Israel, thinks the true explanation may lie not in the matter filling space but in the very laws of physics themselves.

He's come up with a theory called 'modified Newtonian dynamics' (MOND), which argues that the law of gravity deviates at long distances from the standard predictions of Newtonian theory. The model explains phenomena such as spiral galaxy rotation curves (see main text) without the need for dark matter.

In 2004, Jacob Bekenstein, of the Hebrew University of Jerusalem, was able to extend MOND into a form consistent with the principles of Einstein's general relativity. The theory, known as Tensor-Vector-Scalar gravity (TeVeS), is able to explain gravitational lensing observations without the need for dark matter – something MOND on its own struggles with.

More evidence has since emerged on a cosmic scale. In gravitational lensing – where the light from a distant galaxy is bent by the gravity of an intervening cluster (see page 30) – measuring the degree of light-bending makes it possible to infer how much mass resides in the cluster. And when astronomers do this, their findings support the idea that there's far more mass than can actually be seen.

COSMIC DEFICIT

Studies of the cosmic microwave background radiation (CMBR – see page 174) have now provided conclusive evidence. Dark matter is unseen because it only interacts through the force of gravity, while ordinary atoms and molecules, collectively known as 'baryonic' matter, also interact with light through the electromagnetic force. By measuring the effect of electromagnetic interactions on the fine structure in the CMBR, scientists were able to gauge how much of the universe's matter is able to interact this way. Their answer was, once again, 5 per cent.

PARTICLE HUNTERS

Dark matter is very probably made of exotic subatomic particles – of the sort predicted by extensions to the standard model of particle physics (see page 40) such as supersymmetry. It was once thought that 'dark stars' and other large chunks of non-luminous material might contribute, but searches for 'microlensing' events – momentary brightening of a background star through gravitational lensing as a dark body passes in front of it – have drawn a blank.

Various experiments are underway to try and detect dark matter particles, many billions of which could be streaming through your body right now.

In September 2011, scientists from the international Cryogenic Rare Event Search with Superconducting Thermometers (CRESST) collaboration reported 67 particle detections, gathered earlier that year, that could not be explained. A statistical analysis indicated a chance of just 1 in 10,000 that all of these were down to random noise.

Meanwhile, studies published in 2005 by an international team of astrophysicists have affirmed that dark matter was essential for the formation of galaxies – and ultimately ourselves. Galaxies formed by a process called gravitational collapse, where small density irregularities in the early universe grew as their gravity pulled in ever more material. Without the extra gravitational attraction provided by dark matter, this process could not have produced the degree of structure that we see in the universe today.

Vera Rubin (b.1928)

The person who made the decisive breakthrough in the discovery of dark matter was American astronomer Vera Rubin.

She was born in Washington, DC in 1928 to Philip Cooper, an electrical engineer, and Rose Applebaum, who worked for the Bell Telephone Company as a mileage calculator.

Vera completed her BA degree at Vassar College before moving to Cornell to study a masters, where she also met her husband, Robert Rubin. Following this, she took a doctorate at Georgetown University, supervised by eminent cosmologist George Gamow. The subject of her thesis was galaxy clusters – something unheard of at the time, but today well known.

Vera Rubin has four children, all of whom have obtained PhDs in mathematics or physical science. She has received numerous awards, including the Gold Medal of the Royal Astronomical Society and the National Medal of Science.

But recent years have seen an unexpected twist in the story. Dark matter, it turns out, isn't the only bit of the universe that's hidden – some of it comes in the form of an energy field locked away in the structure of empty space. This is the mysterious phenomenon known as 'dark energy'.

The condensed idea
Most of the universe
is invisible

45 Dark energy

In the 1990s, it emerged that the 'dark stuff' which makes up 95 per cent of our universe is, in fact, mostly energy. It was an idea that Einstein had mooted decades earlier, but rejected. Dark energy makes the expansion of space accelerate, profoundly altering the evolution of the universe.

During the latter half of the 20th century, astronomers realized that there was more to the universe than meets the eye – about 20 times more, to be exact. The bright stars and galaxies that fill the night sky represent just 5 per cent of what's really out there. The other 95 per cent is 'dark'. Initially, astronomers thought this dark component was made entirely of matter – most likely exotic subatomic particles (see page 176). In the 1990s, however, evidence emerged that more than two-thirds of it exists in the form of a nebulous energy field pervading space. Because energy and mass are equivalent, this 'dark energy', as it's known, makes a positive contribution to the total mass of the universe.

HUBBLE TROUBLE
In the 1920s, US astronomer Edwin Hubble discovered that the universe is expanding. He studied nearby galaxies and found them all to be moving away from us – and that their speed was, on average, directly proportional to their distance. This is Hubble's law: recession speed equals distance multiplied by a number known as the Hubble constant. What wasn't so clear was exactly how this number is changing with time. All the mass in the universe exerts a gravitational force, which should work

TIMELINE

1933	1980	1998
Zwicky uncovers the first hints that most of the universe is dark	Alan Guth puts forward inflation, driven by a field similar to dark energy	Riess and Schmidt find early evidence that cosmic expansion is accelerating

to gradually slow down the rate of cosmic expansion, causing the Hubble constant to decrease with time.

In 1988, a team led by astrophysicist Saul Perlmutter, of Lawrence Berkeley National Laboratory, California, decided to investigate. They began a study of supernova explosions in distant galaxies. A supernova is a colossal explosion marking the death of a star. They have two very useful properties. Firstly, they are extremely bright, meaning they can be detected in galaxies billions of light years away. This was useful because, in order to track the evolution of the Hubble constant through time, the team had to see galaxies moving as they were billions of years ago. The finite speed of light let them do this – by looking at galaxies a very long way away.

Secondly, supernovae of a particular variety known as type Ia all generate roughly the same amount of light. So if a type Ia supernova is detected in a distant galaxy, then its apparent brightness as measured from Earth can tell astronomers how much its light has been dimmed with distance, enabling them to calculate how far away it is. Combining this information with its speed of recession from us – which can be calculated directly from a measurement of how much its light has been red-shifted (see box) – allows astronomers to work out the Hubble constant for that particular galaxy.

Red shift

Key to inferring the existence of dark energy is the ability to determine how the rate of cosmic expansion has changed between the Big Bang and the present day, by studying the recession speeds of distant galaxies. This is done using a phenomenon astronomers call red shift. When they look at the spectrum of light from a nearby galaxy, it contains a characteristic pattern of peaks and troughs caused by chemical elements in the galaxy absorbing and emitting light at particular wavelengths.

When astronomers look at a faraway galaxy they see the same pattern, only shifted to longer (redder) wavelengths. The shift is caused by the expansion of the universe stretching the light out. If astronomers ever looked at a galaxy moving towards us they would see its light shifted to shorter wavelengths: a 'blue shift'. Astronomers define a measure of red shift as the shift in wavelength of a feature in the light's spectrum, divided by its unred-shifted wavelength. This can then be multiplied by the speed of light to get the galaxy's recession speed away from us.

1999
Perlmutter *et al.* confirm the findings of Riess and Schmidt

2011
Perlmutter, Riess and Schmidt win the Nobel Prize for discovering dark energy

2013
Exact fractions of ordinary matter, dark matter and dark energy are determined

Saul Perlmutter (b.1959)

Perlmutter was born into academia. His father, Daniel, was a professor of bioengineering at the University of Pennsylvania, while his mother, Felice, was a professor at Temple University's School of Social Administration.

Raised as one of three children in the Mount Airy district of Philadelphia, Perlmutter first studied physics at Harvard, before obtaining his PhD – on techniques to search for hidden companion stars to the Sun – from the University of California at Berkeley, in 1986. It was physicist Luis Alvarez who suggested that the same techniques could be used to detect supernovae in distant galaxies, pointing Perlmutter to the work that would ultimately reveal the presence of dark energy.

When Perlmutter did this for a number of type Ia supernovae, he found that the Hubble constant wasn't decreasing with time at all – it was getting bigger. In complete contradiction to what he expected, it seemed the expansion of the universe is accelerating. The team's findings were announced to the world at a meeting of the American Astronomical Society in 1999. In fact, another team led by American Adam Riess and Australian Brian Schmidt had arrived at similar results in 1998. Perlmutter, however, was acknowledged as the driving force behind the discovery, having started work on the project six years ahead of Riess and Schmidt.

Making the universe accelerate requires the presence of some kind of energy that can exert negative pressure to counteract gravity (remember, general relativity asserts that pressure generates gravity, meaning negative pressure generates negative gravity – see page 29). The phenomenon was christened 'dark energy' by American astronomer Michael Turner, in a nod to dark matter. To overcome the attractive gravity of visible and dark matter, dark energy has to make up around two-thirds of the total mass-energy of the universe.

The discovery was voted breakthrough of the year by *Science* magazine, and in 2011 Perlmutter, Riess and Schmidt shared the Nobel Prize in Physics. In 2013, the European Space Agency's Planck spacecraft made detailed measurements of the CMBR – the relic radiation left over from the Big Bang (see page 172) – confirming that our universe is composed of roughly 4.9 per cent ordinary matter, 26.8 per cent dark matter and a whopping 68.3 per cent dark energy.

WHAT IS IT?

When it comes to explaining what dark energy is actually made of, there are two possibilities. The first is that it is built into the fabric of spacetime.

Einstein himself considered such a possibility, before the expansion of the universe was discovered – adding what he called a 'cosmological constant' to the equations of general relativity. In his model, the repulsive force of the cosmological constant exactly balanced the attractive force of gravity on large scales to hold the universe static. However, when the expansion of the universe was later discovered, Einstein retracted the cosmological constant, famously declaring it his biggest blunder.

LOOK AT DR SAUL PERLMUTTER UP THERE, CLUTCHING THAT NOBEL PRIZE. WHAT'S THE MATTER SAUL, YOU AFRAID SOMEBODY'S GOING TO STEAL IT? LIKE YOU STOLE EINSTEIN'S COSMOLOGICAL CONSTANT?
Sheldon Cooper,
The Big Bang Theory

The other possibility is that dark energy arises, not from the structure of spacetime itself but from the materials filling it. A theory called 'quintessence' says that space is covered with a thinly spread exotic matter field that has just the required negative pressure. It's not such a fanciful idea – a very similar kind of field in the early universe probably drove the phase of rapid expansion known as inflation (see page 174). A cosmological constant would be distributed through space perfectly evenly, whereas it's expected quintessence would exhibit at least some degree of clumping – something astronomers should in principle be able to measure. Currently, however, there is no evidence to rule out either theory.

Dark energy has already had a profound effect on models for the evolution of the cosmos, and of the formation of galaxies and clusters. It will continue to do so – impacting especially on theories for the ultimate demise of the universe, as we'll see.

The condensed idea
Space is filled with anti-gravitating energy

46 The death of the universe

Nothing lasts forever, and our universe is no exception. Scientists have combined astronomical observations with our best understanding of gravity and particle physics to deduce how the final days of the cosmos will play out. The good news is it'll be several trillion years before we need to start worrying.

In 1915, Albert Einstein's general theory of relativity let scientists do something they had never done before: build mathematical models of the universe at large. Previously, Newton's theory of gravity had formed the basis for models of simple systems, like planets orbiting their star. But it was of little use for complex systems on larger scales. Einstein's theory offered a solution.

In 1927, Belgian astronomer Georges Lemaître used general relativity to build the first cosmological models. These predicted that space is expanding, as confirmed a few years later by the American astronomers Edwin Hubble and Milton Humason. This fact enabled Lemaître to posit the existence of a beginning to the universe: the Big Bang (see page 172). It also let cosmologists ask if the universe will live forever and, if not, how exactly it will die.

CHOOSE YOUR FATE

There are three main possibilities, depending on the universe's mass content. In the first case, the universe has a lot of mass, causing it to expand

TIMELINE

1927	1997	1999
Georges Lemaître builds the first mathematical models of the universe	Adams and Laughlin speculate on how the universe might end	Astronomers find that nearly 70 per cent of the universe is dark energy

to a maximum size before the gravity of its material content turns the expansion around. Red shifts become blue shifts (see page 181) as space contracts and galaxies move towards each other, ultimately colliding and overlapping. All the matter in the universe is packed into an ever-shrinking volume, and finally crushed out of existence in a colossal antithesis of the Big Bang, known as the Big Crunch. Some theories suggest that the universe could bounce back from a Big Crunch, rising phoenix-like from the ashes in an infinite cycle of expansion and contraction. A universe that recollapses like this is described as 'closed', because its gravity is sufficient to curve space back round on itself to form a closed sphere. Parallel lines in such a universe will eventually cross.

FOR THE FIRST TIME IN ITS LIFE, THE UNIVERSE WILL BE PERMANENT AND UNCHANGING. ENTROPY FINALLY STOPS INCREASING BECAUSE THE COSMOS CANNOT GET ANY MORE DISORDERED. NOTHING HAPPENS, AND IT KEEPS NOT HAPPENING, FOREVER.

Professor Brian Cox

DARK AGE

The second possibility is that the mass of the universe is insufficient to halt the expansion of space. In this case, the universe will continue to expand forever. The gas needed for star formation will eventually be used up, and one by one, stars and galaxies will begin to wink out. Thus begins an eternal cosmic dark age: a thin smear of matter is all that remains of the once-grand universe, being diluted ever further by the relentless expansion of space. Black holes are the final bastions of mass, but eventually even these evaporate away to nothing through the process of Hawking radiation (see page 191). Finally, any remaining atoms decay into fundamental particles – a prediction of grand unified theories of particle physics (see page 48) – and at this point the universe is truly dead. This is sometimes called the 'heat death' scenario for the fate of the universe, because all variations in temperature – which are essential for any thermodynamic processes to take place – have been erased.

Space in this case is not a closed sphere, but is instead infinite in extent. This is called an 'open' universe. It's curved into a saddle shape, rather like

2003 — Robert Caldwell proposes the Big Rip scenario for the end of the universe

2012 — Astronomers confirm the Andromeda galaxy is on collision course with the Milky Way

2012 — NASA's WMAP spacecraft confirms that our universe is flat

A dying universe

In 1997, American physicists Fred Adams and Gregory Laughlin published a 57-page technical paper setting out how the demise of our universe would most likely unfold. Some of the key events are listed in the table below.

5.4 billion years	Our Sun begins to turn into a red giant star, ending life on Earth
6 billion years	Our Milky Way galaxy merges with the nearby Andromeda galaxy
100 billion to a trillion years	All the galaxies in our local group merge; those beyond the local group are no longer visible, swept away by cosmic expansion
100 trillion years	Star formation starts to switch off
1,000 trillion years	Orbits of the last planetary systems decay or are disrupted
10 million trillion to 100 million trillion years	The remains of the last stars are either ejected by their host galaxies or consumed by black holes
10 billion trillion trillion years	Nucleon decay – particles inside the nuclei of atoms begin to break apart
10,000 trillion trillion trillion years	Nucleon decay comes to an end. The universe is mainly composed of black holes, evaporating by Hawking radiation

a Pringle potato crisp. Parallel lines never meet in this universe, but they don't stay parallel either – instead diverging away from one another.

In between these two options is a third possible end state, in which the universe has exactly the right amount of matter so that it will just avoid recollapsing in a Big Crunch but not so little as to become open. In this scenario, the universe is again infinite and will continue to expand forever. But the geometry of space now has no curvature at all, meaning that parallel lines will always remain parallel. That's why this possibility is known simply as a 'flat' universe.

LET RIP

The discovery of dark energy, now known to make up almost 70 per cent of our universe's mass-energy (see page 180), adds a repulsive force to these models, making the open and flat universes expand faster, and perhaps preventing a closed universe from collapsing to a Big Crunch. But in 2003, cosmologist Robert Caldwell, of Dartmouth College, New Hampshire, suggested that dark energy could lead to a fourth possible scenario for the death of the universe. He showed how a particularly extreme form of dark energy, which he called 'phantom energy', could make the universe expand so fast that all the matter within it would literally be torn apart. Called the 'Big Rip', this scenario is only possible if dark energy is due to 'quintessence' (see page 183) – although it couldn't happen for many billions of years.

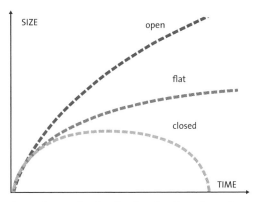

Depending on the density of matter, the universe is deemed to be either open, closed or flat. Astronomical observations favour the latter possibility.

NEW BEGINNING

The best measurements today, taken by NASA's Wilkinson Microwave Anisotropy Probe and released in 2012, suggest that our universe is almost exactly flat, and so is destined to expand forever and slowly fade away. There is some good news, however. First and foremost, the heat death isn't likely for many trillions of years. But even when the time comes, the reality might not be quite so dismal. Models of the early universe hold that a phenomenon called inflation caused a brief period of extremely rapid expansion shortly after the Big Bang (see page 174). And one version of that theory, called 'eternal inflation', says that the rapid expansion is still taking place today – we just happen to live in a small enclave where it's stopped. In this theory, new universes are continually being created as inflation switches off in these regions, too. So although our universe will have come to an end, we can take heart from the fact that somewhere else, across the gulf of space, cosmic life may go on.

The condensed idea
The end is not yet nigh

47 Black holes

Black holes are objects whose gravity is so strong not even light can escape. Anything falling in is doomed to be crushed out of existence at the hole's infinitely dense core. Astronomers can infer the presence of a black hole from the behaviour of nearby bright matter – and many are now known.

Black holes are often viewed as a very modern concept – the stuff of science fiction movies and TV documentaries. Of Stephen Hawking and NASA. But, in fact, the first recorded description of what we now call a black hole was made not this century, or even in the century before. In 1784, English philosopher John Michell investigated what happens to a ray of light in the presence of a gravitational field as described by Newton's theory (see page 12). Newtonian gravity gives rise to a concept known as escape velocity: the faster a projectile is fired into the air, the higher it will get before falling back down, and if you fire it fast enough it will escape Earth's gravity altogether. For the Earth, this speed is 11.2 kilometres (7 miles) per second, but Michell imagined an object with such intense gravity that its escape velocity exceeded the speed of light. He christened it a 'dark star'.

Newtonian gravity, however, could only offer a partial explanation. The true picture would require a bigger and better theory that accounted correctly for the behaviour of light. This arrived in the early 20th century in the form of Einstein's general theory of relativity (see page 28), which was able to describe physics in the presence of the strongest gravitational fields imaginable. In 1916, German physicist Karl Schwarzschild replaced Newtonian gravity with

TIMELINE

1784	1916	1939
John Michell uses Newtonian gravity to propose the existence of 'dark stars'	Karl Schwarzschild formulates the first black hole model based on general relativity	Oppenheimer and Volkoff calculate the maximum mass of a neutron star

relativity in Michell's analysis. He found that if a star – or any object – is squashed below a certain radius then nothing, not even light, can escape its gravitational pull. This 'Schwarzschild radius' for the Sun is about 3 kilometres (1.9 miles); for the Earth it's 1 centimetre (0.4 in).

REALITY CHECK

But could these mysterious objects actually exist in the real world? Many physicists were sceptical, including Einstein. Could an object in space, such as a star, really collapse down to its Schwarzschild radius under its own weight? In a normal star, pressure prevents this from happening. Nuclear reactions in the core of the star generate heat, which creates an outward pressure to balance the weight of the star pressing in.

However, when the star runs out of fuel, nuclear reactions switch off, the pressure vanishes and it begins to collapse. In 1930, Indian astrophysicist Subrahmanyan Chandrasekhar found that as the star shrinks, electron particles are forced closer and closer together. Swiss quantum physicist Wolfgang Pauli had shown that electrons, and in fact all particles in the family known as 'fermions' (see page 40), obey a so-called 'exclusion principle' that prevents them getting too close together. Chandrasekhar calculated that repulsion between electrons would prevent objects weighing less than about 1.4 times the mass of our Sun becoming black holes. Instead, they become what are known as 'white dwarf' stars – typically packing the mass of the Sun into a sphere about the same size as the Earth. Anything heavier, however, would carry on collapsing.

Eventually, positively charged protons and negatively charged electrons are squashed together to form a ball of uncharged neutron particles, (also supported by the exclusion principle), with a diameter of about 10

> **IT'S A PITY THAT NOBODY HAS FOUND AN EXPLODING BLACK HOLE. IF THEY HAD, I WOULD HAVE WON A NOBEL PRIZE.**
> Stephen Hawking

1964
The first black hole, X-ray source Cygnus X-1, is detected by astronomers

1967
John Wheeler coins the term 'black hole' during a talk for NASA

1974
Stephen Hawking discovers that black holes actually radiate particles

Stephen Hawking (b.1942)

Stephen William Hawking was born on 8 January 1942. His father, Frank, was a medical scientist, specializing in tropical diseases. His mother, Isobel, was a secretary. In 1959, he went to Oxford University to read physics. After gaining a first-class degree he moved to Cambridge in 1962 to pursue a doctorate in cosmology.

During his final year at Oxford his movements had started to become clumsy. After tests, he was diagnosed with amytrophic lateral sclerosis, a brain disease that leads to gradual paralysis and death. In 1963, Hawking was given two years to live. Against the odds, and using various technological aids, Hawking celebrated his 70th birthday in 2012. He served as Lucasian Professor of Mathematics (the chair formerly held by Newton) and has been honoured with practically every award in physics short of the Nobel Prize.

kilometres (6 miles) – a so-called neutron star. But in 1939, American physicist Robert Oppenheimer and Canadian George Volkoff found that the maximum mass of a neutron star is around three times the mass of the Sun.

EVENT HORIZON

Anything heavier than this will collapse down past its Schwarzschild radius to form a point of infinite density, a 'singularity'. In 1958, the sphere formed by this radius, marking the point at which the escape velocity exceeds light speed, was named the 'event horizon' by American physicist David Finkelstein. A few years later, physicist John Wheeler, of Princeton University, coined the term 'black hole' to describe such objects. Anything falling through the event horizon of a black hole can never return and is doomed to be crushed to zero size at the singularity.

The 1960s were a golden age for black hole research, with new breakthroughs in our understanding of these mysterious objects flowing thick and fast. In 1963, New Zealander Roy Kerr extended Schwarzschild's maths to describe spinning black holes. Intriguingly, the singularity at the centre of a Kerr black hole is ring-shaped, and some physicists have speculated that a traveller might pass through this to emerge into a new area of space – perhaps a new universe altogether.

SEEING IS BELIEVING

In 1964, astronomers discovered the first candidate black hole – a bright X-ray source called Cygnus X-1. Measurements of the object's mass and maximum size (inferred from its brightness variations) implied that it was almost certainly smaller than its Schwarzschild radius.

During the early 1970s, Cambridge physicist Stephen Hawking and others analysed black holes using quantum theory. Their research revealed fascinating parallels between black hole physics and thermodynamics (see page 20), with four key laws of 'black hole mechanics' emerging that closely mirrored those of thermodynamics. Hawking's research in this area led to the prediction that black holes radiate particles (see box, right).

Today, alongside the many black holes known to have formed from collapsed stars, it's also clear that most galaxies, including our Milky Way, have huge 'supermassive' black holes at their centres. The heaviest-known black hole lies in the heart of the distant quasar S5 0014+81 and truly breaks the scales – weighing in at an astonishing 40 billion times the mass of the Sun.

Black holes ain't so black

In 1974, Stephen Hawking showed that black holes – previously viewed exclusively as voracious cosmic eating machines – can actually lose mass.

The Heisenberg uncertainty principle (see page 37) predicts that space is filled with pairs of virtual particles popping in and out of existence over very short timescales. Hawking considered the case of virtual particles appearing just outside the black hole's event horizon. He reasoned that now and again one particle in the pair would get pulled over the event horizon, while its partner would escape.

The mass of the particle pair is 'borrowed' from the black hole and is normally returned when they disappear an instant later. However, in Hawking's scenario the escaping particle is forced to become 'real', and therefore carries mass away. Over long enough periods, if this 'Hawking radiation' removes more mass than falls in, the black hole will eventually evaporate and disappear.

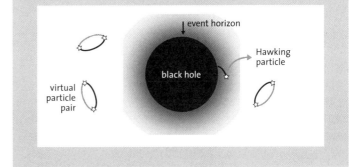

The condensed idea
Gravity so strong, not even light can escape

48 The multiverse

Just as there are many planets, stars and galaxies in our universe, so there is a growing school of thought among physicists that our universe is just one of many – perhaps an infinite number – in a vast, sprawling multiverse. The multiverse is the sum of everything that exists – both the visible and beyond.

What if, out there somewhere, there is a universe in which the Nazis won the Second World War, or humans never evolved, or where you were the first person to set foot on the Moon? The multiverse theory, in all its glory, suggests that there exist universes in which all possible eventualities play out. The idea was first put on a solid scientific footing in 1957, by American physicist Hugh Everett III, working at Princeton University. Everett was schooled in the methods developed by pioneering quantum physicist Richard Feynman. Feynman developed the 'path integral' formulation of quantum theory (see page 38) where the future state of a particle is given by summing over all possible prior states, weighted by their probability of occurring. He called it the 'sum over histories' approach, but never intended these alternate histories to be real. Everett, however, had other ideas.

Everett's theory is called the 'many worlds interpretation' (MWI). According to this view, every quantum event that takes place causes the universe to split into a number of new universes, in which every possible outcome of the event takes place. It offered a new way of viewing the probabilities that the mathematics of quantum theory spits out. For example, if the maths says that a particle has a 60 per cent probability of ending up at location A and a 40 per

TIMELINE

1957	1960	1985
Everett proposes the many worlds interpretation of quantum theory	The term 'multiverse' is coined by Andy Nimmo to describe Everett's theory	David Deutsch lays the foundations for quantum computers, which rely on the multiverse

cent probability of being at location B, then the MWI interprets this to mean that in 60 per cent of universes the particle finishes at A and in the remaining 40 per cent it finishes at B. Given the vast number of quantum particles making up our universe (around 10^{80} – that's a 1 followed by 80 zeros), this means the number of possible universes implied by the MWI is stupendously large.

IN AN INFINITE MULTIVERSE, THERE IS NO SUCH THING AS FICTION.
Scott Adsit, actor

The name 'multiverse', referring to a huge number of interleaved universes (it had already been used to describe a multitude of planets), was coined by Andy Nimmo, of the Scottish Branch of the British Interplanetary Society, in a talk he gave to the society about Everett's work in 1960. The term 'parallel universe' has also entered popular parlance to describe the universes from which the multiverse is comprised.

QUANTUM CLARITY

The MWI explains some of the strange phenomena of quantum theory – such as particles turning into waves that can overlap and influence each other – in terms of interference between nearby universes. In earlier chapters, we saw how the transition from quantum to non-quantum behaviour – a process called 'decoherence' (see page 34) – happens when a delicate quantum state is disrupted by interactions with its surroundings. In the MWI, decoherence can be thought of as parallel universes peeling apart from one another, so that interference between them is no longer possible.

The multiverse view of reality also solves a niggling problem about our own existence. Physicists find that the constants of nature in our universe – such as the relative strengths of gravity and electromagnetism, the mass density of the universe, the amount of dark energy, and the dimensionality of space and time – are all exceptionally finely tuned to the values required for life 'as we know it' to emerge. The values of these constants were determined during

1998
Max Tegmark presents the quantum suicide experiment in its current form

1999
Martin Rees popularizes the view that the multiverse explains cosmic fine-tuning

2003
In a *Scientific American* article, Tegmark proposes a four-level multiverse model

Hugh Everett III (1930–82)

Everett was born in Washington, DC on 11 November 1930. After graduating in Chemical Engineering from the Catholic University of America in 1953, he moved to Princeton to study for a PhD in physics, under the supervision of John Wheeler. It was during this time that he developed the many worlds interpretation of quantum theory. However, it was many years before the theory was taken seriously. In the meantime, a disheartened Everett left academia for industry soon after completing his PhD.

It was while at Princeton that Everett met his wife, Nancy. They had two children together: Elizabeth and Mark. Mark is now lead vocalist and songwriter with the rock band Eels. A heavy smoker and drinker, who took little exercise, Hugh Everett died of a heart attack on 19 July 1982, aged 52.

the early stages of the Big Bang, essentially at random by processes such as spontaneous symmetry breaking (see page 174). In our one universe, why should it be that these numbers take the precise values required to facilitate our own existence? The multiverse offers an explanation. In all of its billions upon billions of parallel universes there are bound to be some in which the constants take the values they do in our own. And it's no surprise that we should then find ourselves living in such a universe.

MULTIVERSE MAP

In 2003, physicist Max Tegmark, of the Massachusetts Institute of Technology, argued that several different levels of multiverse can exist. He proposed a four-tier picture, in which Level I is the universe beyond the furthest distance we can see, given by the distance light has travelled since the Big Bang. The universe can host a large number of such regions this size – an infinite number if it is 'flat' or 'open' (see page 185) – effectively creating a multitude of other universes.

In Level II, the universe is undergoing inflation. This is a period of superfast expansion which is thought to have occurred very early in cosmic history (see page 174). Except that in this scenario, inflation never ended and space is still rushing apart at breakneck speed. Our observable universe just happens to occupy a pocket of normally expanding space, where inflation has switched off. It's believed that such an 'eternally inflating' universe will be constantly giving birth to new baby universes with different constants of nature.

Level III is Everett's many worlds view of quantum theory, while Level IV is the highest tier. Whereas the other levels entertain differences in the physical constants of nature, Level IV supposes that there are universes where the laws of physics themselves can obey fundamentally different mathematical laws. Thankfully, this is as weird as it gets – there is no Level V.

PROOF POSITIVE?

Nevertheless, the very existence of the multiverse still divides physicists. The primary criticism is that it's difficult to test, making the question of its existence an unscientific one. Tegmark, however, believes there is an experiment that could prove it one way or the other (see box, right).

Some have even suggested that the multiverse view of reality could permit time travel – circumventing the usual 'granny paradox', where you go back in time and kill your granny, thus preventing your own existence. But if the universe you go back to is a parallel one, distinct from the one you left, then all is fine. Perhaps the arrival of tourists from the future would provide the ultimate evidence.

Quantum suicide

Some scientists have criticized the many worlds interpretation of quantum theory as untestable. In 1998, Max Tegmark, of the Massachusetts Institute of Technology, refined an idea known as 'quantum suicide' and showed how it could form the basis of a test for the theory.

In the experiment, a quantum event decides whether a gun fires a blank or a live bullet at the experimenter. When the gun fires, the multiverse splits – in one set of universes the experimenter dies, and in the others they live. The experimenter's consciousness must, by definition, always end up in a universe in which they survive. So that's exactly what they see. As the experimenter stares down the barrel, the gun fires blank after blank after blank – if many worlds is correct. However, for those unfortunate enough to witness the test, it's only a matter of time before the experimenter meets their fate.

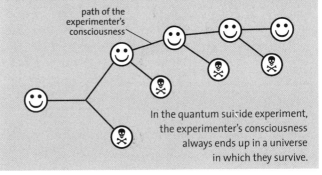

path of the experimenter's consciousness

In the quantum suicide experiment, the experimenter's consciousness always ends up in a universe in which they survive.

The condensed idea
Our universe is one of many

49 Exoplanets

Vast, powerful telescopes are searching areas of the Milky Way for exoplanets. These are planets that orbit stars outside our own Solar System. Thousands have been discovered since 1995 and scientists now believe there could be hundreds of millions of Earth-like planets out there with the potential to sustain life.

In July 2015, the scientific community was buzzing with excitement following a discovery by NASA's Kepler Space Telescope. The mission was launched to find Earth-like planets outside our Solar System, orbiting stars similar to our own, and it did just that with the exoplanet Kepler-186f, some 500 light years from Earth in the constellation of Cygnus. It orbits a cool dwarf star and, vitally, is in the star's 'habitable zone', where water could exist on its surface in liquid form.

The mission has found thousands of planets since its launch, but none as similar to our own as Kepler-186f. The planet is slightly larger than Earth, with a diameter just 10 per cent bigger. Previous exoplanets found in habitable zones were at least 40 per cent larger than Earth. These factors combine to make it one of the most intriguing extrasolar planets ever seen.

DISCOVERY OF EXOPLANETS

For centuries, scientists and philosophers suspected that other stars would have planets orbiting them. In 1855, Captain W.S. Jacob of the East India Company's Madras Observatory claimed to have detected a planet orbiting the star 70 Ophiuchi. This was later refuted, however. Further claims were

TIMELINE

1855	1992	1995
W.S. Jacob mistakenly claims to see a planet orbiting 70 Ophiuchi	An exoplanet is found that orbits a pulsar in the constellation of Virgo	Mayor and Queloz discover the first exoplanet, 51 Pegasi b

made from the mid-20th century onwards, and in 1992 astronomers found the first planetary system beyond our own. Unfortunately, however, it was orbiting a pulsar (the dense remnant of a supernova explosion), rather than a still-active star.

Success finally came in 1995 when Michel Mayor and Didier Queloz of the University of Geneva announced they had found an exoplanet orbiting 51 Pegasi, a star similar to the Sun. Mayor and Queloz knew that as a planet orbits, its gravity pulls on its star. They tracked down the 'wobble' created by an orbiting planet by searching for variations in the frequency of the star's light. '51 Pegasi' is 50 light years from Earth in the constellation of Pegasus. Its planet is believed to be around half the size of Jupiter, with a surface temperature of around 1,000°C (1,800°F). Closer to its star than Mercury is to our Sun, it moves at an astonishing speed, completing an orbit every 4.2 days. Formally designated 51 Pegasi b, the planet is also known as Bellerophon after the Greek hero who tamed Pegasus, the winged horse.

> **ONE IN 200 STARS HAS HABITABLE EARTH-LIKE PLANETS SURROUNDING IT – IN THE GALAXY, HALF A BILLION STARS HAVE EARTH-LIKE PLANETS GOING AROUND THEM – THAT'S HUGE, HALF A BILLION. SO WHEN WE LOOK AT THE NIGHT SKY, IT MAKES SENSE THAT SOMEONE IS LOOKING BACK AT US.**
> Michio Kaku

Following this landmark discovery, more exoplanets were located using Mayor and Queloz's 'wobble' method. Those similar to 51 Pegasi b became known as 'hot Jupiters', due to their size and conditions. The fact that these planets orbit close to their stars seemed paradoxical, however, because at the time it was believed that giant planets should be much further away. Astronomers now believe that friction from dust and gas has eroded the orbits of these planets, drawing them closer to their stars.

Astronomers also began to discover exoplanets of different kinds, such as massive ice planets and 'super-hot Earths'. Solar systems like Upsilon Andromedae revealed multiple planets in orbit. But the race was on to find an exoplanet similar to our own.

2009	2015	2015
NASA launches the Kepler space telescope to detect exoplanets	Kepler-186f is found – an Earth-like planet in its star's habitable zone	Kepler finds GJ 1132b, a 'twin Venus', just 39 light years away

The Kepler mission

The Kepler space telescope orbits the Sun every 371 days. It is 4.7 metres (15.4 ft) in length and weighs 1,052 kilograms (2,314 lb). Its main sensor, a photometer, constantly measures the brightness of 100,000 stars in one area of sky, across the constellations of Cygnus and Lyra. Starlight is funnelled directly from Kepler's 1.4-metre (56-in) mirror to a 95-megapixel camera, which can capture tiny 'winks' in brightness as small as 20 parts per million. The telescope has an extraordinarily large field of view – 105 square degrees.

By January 2015, Kepler had discovered 1,013 exoplanets, with around 3,200 more awaiting confirmation. The mission has not been without glitches. Two of the four reaction wheels used for pointing the spacecraft failed in 2013 and 2014, but the craft is still functioning in a reduced capacity.

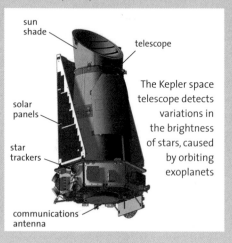

sun shade

telescope

solar panels

star trackers

communications antenna

The Kepler space telescope detects variations in the brightness of stars, caused by orbiting exoplanets

PLANET HUNTING

In 2009, NASA launched the Kepler space mission to do just that. Named after the 17th-century German who discovered the elliptical orbits of the planets (see page 167), Kepler uses the 'transit' method to detect tiny dips in the brightness of a star when a planet passes before it. In conjunction with the wobble method, this allows astrophysicists to calculate the planet's size, mass, temperature and orbit.

Spectroscopy can be used to deduce the composition of an exoplanet's atmosphere. When the planet orbits in front of its star, starlight passes through its atmosphere, boosting or weakening the brightness of certain colours as the light is absorbed and re-emitted by the particular set of atmospheric chemicals. As the planet moves behind the star, however, only the light from the star can be seen. The difference between the two signals allows the spectrum of the exoplanet to be calculated, yielding details of its atmospheric composition.

TWIN VENUS

Fresh excitement was sparked in November 2015 with the discovery of exoplanet GJ 1132b in the constellation of Vela. This was detected by the MEarth-South array, a group of eight robotic telescopes at the Cerro Tololo Inter-American Observatory in Chile. MEarth-South can detect transits among red dwarf stars 100 light years from Earth. Follow-up observations with the HARPS (High Accuracy Radial velocity Planet Searcher) spectrograph, on the 3.6-metre (11.8-ft) telescope at La Silla Observatory, also in Chile, found the planet to have 1.6 times the mass of Earth. As the planet races around its star, orbiting every 1.6 days, it produces a 0.3 per cent dip in starlight, which was detected by the telescopes. GJ 1132b has a diameter of around 14,800 kilometres (9,200 miles), making it 16 per cent larger than Earth. Its oven-like temperatures are too hot for life, yet potentially cool enough to maintain an atmosphere. Some scientists are calling the planet a 'twin Venus'. At a relatively nearby 39 light years, it's also a lot closer to our Solar System than Kepler-186f, making it perfectly placed for further study from Earth.

Gravitational microlensing

Detecting giant exoplanets that, like Neptune and Uranus, are a great distance from their star is extremely difficult. In 2014, astronomers discovered such a planet in a binary star system, 25,000 light years from Earth, using the Warsaw Telescope at Las Campanas Observatory, Chile. The planet is around four times the size of Uranus, and orbits its star at around the same distance that Uranus orbits our Sun.

It was discovered through a technique called gravitational microlensing. The gravity of a star can focus the light from a distant star and magnify it like a lens. Occasionally, the presence of a planet orbiting the lens star can be detected in the magnified light signal. Some astronomers believe that microlensing could detect other planets in very wide orbits. These chance alignments between objects are extremely rare, however, occurring once in a million years for a given planet.

Increasingly powerful and sophisticated telescopes are now under construction, such as the Giant Magellan Telescope in Chile, which will have a resolution ten times sharper than the Hubble Space Telescope. Meanwhile, in 2024, the European Space Agency plans to launch PLATO (Planetary Transits and Oscillations of stars), a planet-hunting space mission that will concentrate on exoplanets orbiting in the habitable zone around Sun-like stars. These exciting new projects may well reveal the first signs of life on worlds beyond our Solar System.

The condensed idea
Planets orbiting distant stars

50 Extraterrestrial life

The discovery of extraterrestrial life has left the realm of science fiction and is tipped to become a reality within the next decade. NASA aims to land astronauts on the planet Mars by 2040 to search for alien life forms, and powerful telescopes are scanning the skies for radio signals emitted by intelligent extraterrestrials.

The universe is teeming with alien life. This is the consensus of leading astrobiologists, who say they now know where to look and how to search for extraterrestrials. Astrobiology is the term used for the study of life elsewhere in the universe, and there's a wide range of possibilities: some planets may seem devoid of life, but hide a range of fossils and organic matter; others may host simple organisms such as bacteria and viruses; and there may be intelligent alien life out there with the technology to communicate with us.

NASA estimates that it will find signs of life in our Solar System by 2025 and be able to verify it within 20 to 30 years. Life on Earth originated in the oceans, so the hunt for water is key in the search for life elsewhere. In 2015, NASA found strong evidence for the presence of water on Mars's surface. This was indicated by streaks left by hydrated salts that sporadically seep to the planet's surface. The European Space Agency's ExoMars rover is scheduled to launch in 2018, carrying a variety of instruments specifically designed to search for life, while NASA hopes to land astronauts on Mars in the 2030s.

TIMELINE

1959
Morrison and Cocconi publish first research paper on SETI

1977
The inexplicable Wow! Signal is detected from deep space

1989
Galileo is launched to examine the planet Jupiter and its moons

The 1990s Galileo space probe to Jupiter found strong evidence of a frozen ocean under the crust of Europa, one of the Jovian moons. NASA believes that Europa has all the necessary ingredients to support simple organisms – a rocky seabed, salt water and energy driven by tidal heating. A further reconnaissance is planned for the 2020s.

> **WE'RE GOING TO HAVE STRONG INDICATIONS OF LIFE BEYOND EARTH WITHIN A DECADE.**
> Ellen Stofan,
> NASA Chief Scientist

OUTER SPACE

Since the late 19th century, mankind has been fascinated with the concept of intelligent alien life. In 1901, the Serbian inventor Nikola Tesla announced that he had received signals from alien life forms in our Solar System via wireless electrical transmissions. Although considered eccentric by some, Tesla may have set a precedent for the modern use of radio telescopes in the hunt for aliens. Since the 1950s, SETI (the Search for ExtraTerrestrial Intelligence) has gathered together resources and brains from across continents to search for messages from space in the form of radio waves, using telescopes designed to look for natural cosmic radio sources.

The Wow! Signal, detected in 1977, was a radio signal picked up from space that could not be explained by any known phenomenon. The letters and numbers each correspond to signal intensity in a 12-second interval. Those circled are exceptionally high.

THE WOW! SIGNAL

For a long time nothing was detected apart from the background hum of space. Then, on 15 August 1977, American astronomer Jerry R. Ehman noticed an unusually sharp and distinct narrowband radio signal picked up by the Big Ear telescope in Ohio. The signal was 30 times louder than anything else around it, and appeared to have come from the constellation of Sagittarius. It couldn't be explained by natural sources on Earth or in our Solar System. Ehman circled the numbers on a computer printout and wrote 'Wow!' next to it, hence the signal's name. Astronomers have tried to detect the Wow! Signal

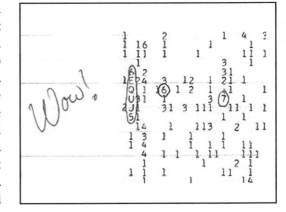

2003

NASA launches its twin Mars Exploration Rovers to the Red Planet

2009

Kepler space telescope begins the search for Earth-like exoplanets

2015

NASA detects evidence for the presence of water on Mars

Sustaining life on other planets

Life on Earth is dependent on a group of elements – primarily carbon, oxygen, hydrogen, phosphorus, sulfur and nitrogen – with water as a solvent for biochemical reactions. Life on other planets could be dependent on a completely different chemical cocktail, but there are cornerstones that scientists agree would be essential for extraterrestrial life to flourish. The main chemical ingredients should be abundant in space and there should be a liquid medium with a wide temperature range for chemical reactions to take place. Any planet supporting alien life would need an atmosphere and some liquid on or near its surface.

The Kepler space telescope, along with other powerful telescopes on the ground, is searching the skies for exoplanets (see page 198) that orbit suns similar to our own. Those found in the 'Goldilocks zone', where conditions are neither too hot nor too cold to prevent water being present in liquid form, may be home to extraterrestrial life. Based on current discoveries, there are half a billion of these planets in the Milky Way.

many times since, but to no avail. Was it a tantalizing glimpse of intelligent extraterrestrial life? To date, no one knows.

Another inexplicable signal was detected in 2010, this time from a relatively nearby galaxy. Radio waves were being transmitted from an unknown object in Messier 82 (M82), which is 12 million light years from Earth in the constellation of Ursa Major. The detection was made by scientists at the UK's Jodrell Bank Observatory, using the MERLIN array of radio telescopes.

This signal was unlike anything astronomers had previously detected – moving fast, with a sideways velocity apparently four times the speed of light. Such superluminal motion is an optical illusion that has previously been detected in material squirted out of supermassive black holes. M82's centre probably contains such a black hole – but the mystery object was not near the galaxy's centre. No other theory fits the data, and this object remains one of astronomy's unsolved puzzles.

ALIEN MEGASTRUCTURE?

In 2015, intriguing anomalies were picked up by the Kepler space telescope (see page 198). NASA revealed that the star KIC8462852, which is 1,500 light years away in the constellation of Cygnus, had inexplicably dimmed by up to 20 per cent several times in recent years. The dimming was far too great to be caused by a planet passing before it, while another possible theory – the break-up of a comet – would result in a detectable excess of infrared radiation, which wasn't seen. More recently, it's been suggested that the dimming could be explained by a cloud of exceptionally cold comets, but these would need to be on a very unusual orbit to explain the observations.

Astronomers were left scratching their heads, and some began to speculate that the dimming was caused by a structure built by intelligent alien life. Such an object would have to be vast in order to cause the light variations registered by Kepler, and as a result, a buzz began to circulate about an alien 'megastructure'. The search for a plausible explanation continues.

Extremophiles

Organisms that can survive in extreme conditions are known as extremophiles. They have become an important area of research for astrobiologists. Microbes have been found thriving around hydrothermal vents on the ocean floor and in ice, or very acid or alkaline environments. Some can even exist in the core of a nuclear reactor. Extremophiles that live in very cold conditions are of great interest to astrobiologists because most of the planets and moons in our Solar System are frozen. Their discovery has expanded the number of possible habitats where extraterrestrial life might be found.

Kepler is locating many more rocky, or Earth-like planets, orbiting in the habitable zone around their stars, raising the likelihood of alien life forms existing throughout the universe. The vastness of space limits our potential to encounter intelligent extraterrestrial life, but it also protects us. Stephen Hawking has warned that, should aliens reach Earth, it's highly likely that they would aim to 'conquer and colonize'. Hawking is keeping an open mind, however, and lending his support to the Breakthrough Initiatives, a project that aims to accelerate the search for intelligent aliens and learn how best to communicate with them. As the late, great cosmologist Carl Sagan put it: 'somewhere, something incredible is waiting to be known.'

The condensed idea
We are not alone

Glossary

Abiogenesis The concept that life on Earth arose from non-organic matter.

Allotropes Different forms of the same element.

Antimatter Most particles of matter have counterpart 'antiparticles'. The antimatter particles have the same mass and spin as ordinary particles, but key properties, such as electric charge, are reversed.

Atom The smallest unit of a chemical element. Atoms consist of a nucleus with protons and neutrons, surrounded by electrons.

Atomic number The number of protons in the nucleus of an atom. Elements on the periodic table are listed in order of their atomic number.

Bacteria Single-celled organisms that are found in almost every environment on Earth.

Baryon A class of subatomic particle that includes neutrons and protons, which make up the nuclei of atoms.

Bequerel The unit of radioactivity of a substance, named after Henri Becquerel.

Bit The fundamental unit of information. Short for 'binary digit', it can take the value either 1 or 0.

Carbon cycle The circulation of carbon from Earth to the atmosphere and back, via plants and animals.

Catalyst A substance that accelerates the rate of a chemical reaction without being changed or consumed. Enzymes are biological catalysts used in cells.

Cell The smallest unit of life. Eukaryotic cells have a nucleus whereas prokaryotic cells do not.

Cosmic microwave background radiation The relic radiation from the Big Bang in which the universe was created.

Dark energy An energy field pervading space that causes the expansion of the universe to accelerate. Dark energy is the dominant component of the universe.

Dark matter Matter that only interacts gravitationally, and so is invisible.

DNA The molecule, found in all cells and many viruses, that carries the organism's unique genetic code.

Element A substance that cannot be broken down into any other substance. Elements are the fundamental building blocks of chemistry.

Entropy The degree of disorder in a thermodynamic system. Also used in information theory to quantify the amount of information in a signal.

Evolution The change in characteristic traits of organisms through the generations. Individuals with traits most suited to their environment are more likely to survive and reproduce.

Fermion A class of subatomic particles with half-integer quantum spin.

Fractal A complex shape that looks the same on all length scales. Fractals govern the dynamics of chaotic systems.

Fullerene Molecules of carbon atoms that are linked together in a sphere (buckyballs) or cylinder (nanotubes).

Galaxy A group of stars that are bound together by gravity. Most are believed to have black holes at their centre.

Genome An organism's complete set of genetic data, including DNA and genes.

Hadron A class of subatomic particles, all of which interact with the strong nuclear force. It includes protons, neutrons, mesons and quarks.

Higgs boson A subatomic particle that gives mass to all the other particles in the standard model.

Inertia The resistance of heavy objects to changes in their current state of motion.

Lepton A class of subatomic particles that don't feel the strong nuclear force. It includes electrons, tauons, muons and neutrinos.

Molecule The smallest possible amount of a chemical compound. Molecules form most of the matter on Earth and consist of two or more atoms bonded together.

Nanotechnology Manipulation of individual atoms to manufacture materials.

Qualia Our perceptions of different conscious experiences in the brain.

Quantum electrodynamics The quantum field theory of the electromagnetic force.

Quantum spin A property of quantum particles describing their symmetry under rotation.

Spacetime Space and time as one unified entity, a concept central to the theory of relativity.

Standard model Our best description of the laws of particle physics.

Superconductor A material that can conduct electricity with zero resistance. They offer the potential to improve electrical power generation, through reducing waste.

Tectonic plates Massive rock plates in Earth's crust that move constantly. They account for continental drift, deep-sea ridges, earthquakes, mountain formation and volcanoes.

Universal grammar The theory that the rules of grammar are hard-wired into the human brain.

Virus The smallest organisms of all, only capable of replicating by injecting their DNA into host cells.

Index

First published in the UK in 2016 by

Quercus Editions Ltd
Carmelite House
50 Victoria Embankment
London EC4Y 0DZ

An Hachette UK company

Copyright © Quercus 2016

Paul Parsons and Gail Dixon have asserted their right
to be identified as authors of this Work.

Design and editorial by Pikaia Imaging

A CIP catalogue record for this book is available
from the British Library

HB ISBN 9781784296148
EBOOK ISBN 9781784296155

10 9 8 7 6 5 4 3 2 1

Printed and bound in China

To our amazing son Callum – without
whom this book would have been
finished in half the time.

Picture credits:

95: Photo by D. Carr and H. Craighead,
Cornell University.

All other pictures by Tim Brown.